OUR FUTURE:
DR MAGNUS PYKE PREDICTS

DR MAGNUS PYKE

Dr Magnus Pyke, OBE, has made science intelligible for the ordinary person. Loved by millions for his lively and stimulating television appearances, he is, nonetheless, a very highly qualified scientist and has held a number of important posts in government departments and industry, most of which were concerned with nutrition. Author of many books, he claims that for recreation he 'writes a page a day and savours the consequences'. Married, with two grown-up children, he lives in Hammersmith, West London.

OUR FUTURE:

DR MAGNUS PYKE PREDICTS

Hamlyn Paperbacks

OUR FUTURE: DR MAGNUS PYKE PREDICTS

ISBN 0 600 20013 2

First published in Great Britain 1980
by Hamlyn Paperbacks

Reprinted 1980

Hamlyn Paperbacks are published by
The Hamlyn Publishing Group Ltd,
Astronaut House,
Feltham,
Middlesex, England

Made and printed in Great Britain by
Hazell Watson & Viney Ltd
Aylesbury, Bucks

CONTENTS

IT'S ONLY FIFTY YEARS

What is fifty years? Over two-thirds of human life? Half a century? Five decades? 2,600 weeks? It is entirely a matter of perspective. Looking back fifty years to 1930 seems a lot closer than looking forward the same amount of time to 2030, a date that lies beyond the boundary of the new century.

The future is a common unknown shared by us all, just as the past is a point of common reference. Our speculations may differ widely on the nature of the future, but events in the past are incontrovertible.

When we look into the future we must be conscious the whole time of the past. We should look to history to guide us into the future, because today's future will be tomorrow's history.

And when we do look to history we see that fifty years is not a long time in the affairs of men. Many of the great inventions of our present age took nearly half a century or more before they made their impact on our lives. Photography was invented fifty-six years before it was practically realized; television sixty-three years; nuclear energy forty-six years and radar thirty-five years.

So looking ahead to the year 2030 means we must look back to 1930 and measure the real changes in our lives since then, to put the changes of the future into perspective.

The Shape of Things to Come

The development of the species

'Only mothers can think of the future – because they give birth to it in their children,' wrote Maxim Gorky in 1910. Nodding in agreement, today we still applaud this noble panegyric to the sacred nature of motherhood. Yet even now as we gaze uncertainly into the crystal ball of futurology to see what lies ahead in the next fifty years, the picture is clouding with sinister predictions, threatening the basis of our time-honoured belief in natural selection and the development of man from one generation to the next. Half-realized images of embryo-transplants, genetic engineers and test-tube babies drift in front of our unwilling eyes; we turn away with disgust and put our faith in tradition and human nature, but the images drift closer all the time sharpening in definition as they approach.

The great scientific accomplishments that have led to our initial attempts in coming to understand genetic structure have also had far-reaching consequences in our ability to control, and even reproduce, the great 'mystery of life', childbirth. We can already determine the sex of a foetus in its early months of development. We can also test it for the presence of harmful genes and abort it if the tests reveal the presence of a serious genetic disorder. Other medical techniques now make artificial insemination a practical alternative for women whose husbands are sterile or carry defective genes. They can receive sperm from a genetically screened, anonymous donor, and then enjoy a perfectly normal pregnancy. Genetic deficiencies can also be identified in a growing embryo fertilized in a test tube, until doctors are satisfied that it is genetically sound, whereupon it is reimplanted in the uterus to grow normally. As a logical extension to this process it should be possible to implant an embryo from a woman incapable of going through a pregnancy into the uterus of one who is. This surrogate mother could then give birth to the child and return it to its natural mother once it is born. It is not necessary to be a Darwin or a Galton to look a little further ahead and see the obvious conclusion of this work – the ability to bypass the natural process of reproduction altogether. If it is possible to fertilize and grow an embryo in a test tube, why should it not be possible to complete the growth there? After all, incubators have been in common use for years at the other end of the reproductive cycle helping to maintain some of the functions of the pregnant mother until the premature child is fully developed. So how long will it be before the two artificial operations can be put together?

Writing a century before 2030 and looking 600 years into the future Aldous Huxley anticipated such a scenario in *Brave New World*. But will Huxley's vision of the mass decanting of pre-conditioned Alphas, Betas, Gammas, Deltas and Epsilons really be 500 years away in 2030? Will the sacrament of marriage still contain the clause about the procreation of children as one of its principal tenets? Will marriage even exist?

Would it not be preferable to store one's eggs or sperm in the flower of youth, become sterilized afterwards and enjoy total sexual freedom and licence without the slightest risk of conception? The 'materials' would always be in cold storage waiting for the time when we would want to raise a family and even then, for a woman, a choice would lie between experiencing parturition herself or delegating the burden to skilled laboratory staff. For those women who wished to become mothers but not wives there would be embryo-transplants available over the counter. These could be carefully arranged to provide a wide permutation of hair colours, eye colours, sex, height and intelligence to correspond to the mother's ideal of the perfect child.

Of course, the principal difference in Huxley's scenario was that this element of choice was removed from the individual. The population was to be carefully controlled, with social categories filled according to their needs. The workforce, for example, was to be drawn from the serried ranks of lower order clones, identical beings duplicated by another process of genetic manipulation. But clones are more than the product of a fertile imagination. It is over ten years since the first experiments with amphibians produced genetically identical tadpoles. The nuclei of a batch of frog's eggs were destroyed by radiation and were replaced by other nuclei taken from the cells of a tadpole's intestine. Once the full set of chromosomes had been restored in each egg they began to divide as normal and eventually produced tadpoles identical to the one from which the 'donor' nuclei had been originally taken. Admittedly work like this is far more difficult to conduct on mammal cells because they are very much smaller, but researchers are confident that similar cloning techniques will be developed for mammals during the next fifty years and at the top of the mammal list is man himself.

So might the individual be no longer responsible for his posterity? Might governments set out to breed super armies of identical combat troops? Might our concepts of the family wither away in the antiseptic, regulated society of tomorrow when love and sex no longer play any part in the reproductive process and babies come in bottles, not on draught?

The answer, fortunately, will be no. Science and technology might take us to the edge of the precipice but in the end we will turn back, recoiling at the prospect of what we might have perpetrated. Brave New Worlds only occur if we allow them to and in some cases we are mindless enough to let them develop, but in this case, when our basic mechanism of survival is in jeopardy, we will rebel. At present it is fashionable to debate the value of marriage. There are working parties busily discussing alternative social nuclei to replace 'wedlock' and the family unit. But this is the bright, prismatic, non-representative view of television thinking, which once upon a time was called 'long-haired'. In reality, the great mass of the population like to get married and have a family. Even the current epidemic of divorces and separations will gradually lead people to appreciate that this course of action is not necessarily an avenue towards happiness. In fifty years the tide will have turned and we will see that a stable relationship is a thing to be desired not shunned.

We are going to see much more clearly then than we do in our present unthinking age, that life is a particularly tricky thing. Man is not God. We may do these clever scientific 'conjuring tricks', but nevertheless we are still susceptible to danger, to accidents, to unexpected death and hardship which we cannot control. Trying to alter the natural course of human development is full of pitfalls. Remember the well-known anecdote about Isadora Duncan and Bernard Shaw? 'Mr Shaw, let us have a child, it could have my looks and your brains,' said the lady. 'All very well,' replied Shaw, 'but suppose it had my looks and your brains?'

The same answer will be given to those who campaign for population control. Human populations are self-conscious entities and the 'population explosion' is a mistaken concept. Children do not come bang from guns like puffed rice. They come because people want them. If an Indian farmer wants seven strong sons he will go on producing children until he has achieved his goal and no virtuous admonitions from you or me will stop him from doing what is, after all, human nature. I suspect that in the end the current Western vogue for single-parent families and for living alone will disappear. Although there are a few who are undeniably suited to these styles of life, there are many more who, I am sure, suffer from loneliness, isolation and unhappiness as a direct result of this domestic exile. In fact, I suspect that all our attempts to construct alternative codes of ethics and conduct will flounder, as in the old rhyme:

There was a young man of Moldavia
Who didn't believe in the Saviour,
But invented instead, with himself at the head,
A system of moral behaviour.

In fifty years we will apprehend that we cannot and should not do that.

The actual size of the family will therefore be conditioned by whatever social circumstances prevail in 2030. But I suspect that we will not go far wrong if we think of a man and a wife and two children, though there will obviously be great deviations either way. There will be people who like children and, if they can afford it, they will have more than two. We will probably continue with a vaguely zero growth population (in the UK), or one which might even show a slight fall, so the average will probably be two children to a family. Then there will be the question of Granny, who will be living longer than she does today. Here, I foresee Granny moving in, because she will be so useful around the house and helping to babysit, while Grandpa, if he survives, will be occupied papering the spare bedroom or looking after the garden. So after the anxieties and experimentation of the last half of the twentieth century the nuclear family will be reinstated as the key unit in future society. The next question is, where will the family be living?

Home, sweet home

One common fact which does emerge from the myriad prognostications and predictions about life in the future is that there will be considerably more of us inhabiting the earth in fifty years' time than there are today. The actual figure is understandably a matter of almost arbitrary guesswork, but if the current growth rate of one and a half million every week continues into the next century, the world population will have more than doubled by 2030. If an additional trend, increasing urbanization, also continues, all these extra people will be flocking to the ever expanding cities to set up their 'Casa Nostras' and 'Chez Nous'.

In 1900 there were nineteen cities in the world with populations of over one million, and of these the largest was London. Only sixty years later the number had risen to 141 and it has been predicted that by the end of the century half the people on earth will be living in urban areas. Mexico City, currently ranked as the world's number one, has a population three times greater than the total population of Norway. But by the turn of the century it is expected to have grown two and a half times, to a staggering total of thirty million inhabitants.

The overriding question posed by these statistics is where will all these future citizens be accommodated? Will the urban planners of the world's city halls be content to allow them to develop vast shanty towns that would make life in the poorest areas of present-day Calcutta seem like a garden party at Buckingham Palace or the White House? As it is, conservative estimates suggest that at least one million people are living on the streets of Calcutta, with no roofs over their heads and no beds but the pavements to sleep on at night.

Might the planners, on the other hand, take bold, imaginative strides and develop space-age cities inside climatically controlled domes or deep in the bowels of Mother Earth? If they do choose this course there is no shortage of plans for them to work from.

The architect Paolo Solari is in the forefront of these futuristic designers. He has proposed the construction of cities capable of accommodating 170,000 people on less than 1·5 square kilometres of land. This city would take the form of two pyramids joined at their bases, one pointed up, the other pointing down. Inside there would be accommodation, leisure areas, places to work and shopping precincts. All the essential services would be contained within the structure and there would be heliports and other communications networks to link one city with the next.

Other designers have looked to the ocean, from which life on land evolved, as the answer to our future overcrowding. They envisage communities living in floating island cities, like enormous deep-sea drilling rigs. These would extend about 100 metres below the surface so that in shallow coastal waters they would probably sit on the bottom. They would provide convenient overflow accommodation for densely populated coastal areas like the eastern seaboard of the USA, and they might even be situated on main shipping lanes across the ocean, to provide stopping-off points for tourists and long-distance travellers.

Whatever designs are adopted in the future, though, the

fact still remains that we have already allowed huge sprawling conurbations to radiate out from our inner-city areas, bringing with them serious social side effects. Part of the reason for this growth in urban areas has been the vast improvement in communications. With the telephone, the motor car and the subway there is no longer any need to live within convenient walking distance of our daily occupations. We can afford the luxury of leading two lives – the working life spent in a totally work-oriented environment like the industrial estate, the City of London or Wall Street, and the domestic life spent miles away in the pleasant suburbs or even further out in the surrounding countryside.

This situation worked ideally fifty years ago when it was possible to step on to the beautiful Metro to be whisked away to downtown Pinner, or Burbank, in less than an hour. But, sadly, it has deteriorated as other people have cottoned on to the same idea. The result today is that while on the one hand the population of inner-city areas is falling, on the other the number of commuters pouring into the cities is rising every year.

At the same time the spread of the city unit and the split that has occurred in most people's lives have led to a growing isolation of the individual. It is not uncommon now not to know the people who live next door. Nor is it uncommon to hear of individuals surrounded by sophisticated means of communication who are alone and incapable of communicating with anyone. The result is dissatisfaction and frustration.

The commuters know that they have backed a loser, because travelling backwards and forwards every day is ruinously expensive and increasingly interrupted by cancellations or delays. Even when they are able to perform this daily ritual they can only do it in intense discomfort. For those who opt to travel by car the picture is equally gloomy. Driving nose-to-tail in a rush-hour traffic jam they clog the roads with virtually stationary vehicles and pollute the air with carbon monoxide. At night the lonely occupant of the city apartment watches them all struggling out again in the certain knowledge that he will be left alone again to listen to the sound of the wind mournfully blowing through the 'concrete jungle'.

However, an additional factor that has only recently come into play may be the trump card needed to win this frustrating urban game. This factor is the rising cost of energy. For if we reach the situation when commuting becomes economically impossible, we shall be forced to live nearer our work, schools and shops. This could form the basis of a strong case for creating sub-units within our large built-up areas, units the size of villages or moderate-sized boroughs, in which people could conduct their whole lives, yet still enjoy the facilities of the large city if they chose to do so.

One of the chief advantages of living in a borough, or a village, is that it provides a unit in which you can feel totally at home. Perhaps it is interesting to note that the original meaning of the word 'home' in Old English was 'a village, town, or a collection of dwellings'.

The London in which I live was originally a group of separate villages that gradually grew together to form a great city, and the identity of the individual villages has survived in residual form in the boroughs that bear their names. I consider myself a Hammersmith man, because I have lived in the borough for a long time and I feel I know it well. I get a pleasant sensation of domestic loyalty when the Hammersmith paper phones me up, and I am equally indignant when people assume that I live in Chiswick, because the people of Chiswick think they are posher than us. But maybe they are just as proud of their own borough.

At present we pat ourselves on the back, saying how clever we have been in building New Towns to ease overcrowding in large cities. But these developments lack the essence of community spirit that characterized the slum areas they were intended to replace. In the future, experience will teach us that cities cannot be created, they need a certain amount of organic growth. Our ancestors in the late eighteenth-century villages were united by an interest in local affairs. Irrespective of their social status they enjoyed a common, unifying bond. Yet within that framework individuals lived their own private lives behind their front doors and drawn curtains.

Naturally it would be a disastrous mistake to try to recreate Jane Austen's England in the twenty-first century, but the underlying strengths of its social units could provide valuable examples of a better environment for living in fifty years time.

'No man is an island, entire unto himself,' wrote the poet John Donne and in the year 2030 we will be more conscious of our fellows than ever before in our history. At the same time, it will be important not to lose our individual identities entirely by becoming too intimately members of one another. Although I depend on the services of a great city like London, I also have to rely to a very large degree on my immediate neighbourhood and my own resources. This is an important moral lesson as well as an ordinary lesson of life, which will assume even greater significance during the next fifty years.

There's No Place Like Home

Four square walls

With the ever-pressing need for accommodation, some architects and designers are re-examining the possibility of 'instant' buildings, 'instant' campuses and 'instant' factories as the solution to future growth, now that tower blocks have acquired a bad reputation. We are already becoming familiar with the portable huts which stand on adjustable legs like patient pack animals, and wait to be filled with all the paraphernalia that will instantly transform them into school classrooms, construction site offices and public lavatories. Until now they have not been used as private dwellings, but according to some projections the era of prefabricated 'instant' homes is just around the corner, if problems like sound insulation can be overcome.

If these projections are right we can look forward to beehive cities in the next century, formed from clusters of self-contained units arranged in asymmetrical piles. Each unit will be set in a steel frame that will give it rigidity and strength. These frames will be used to lock the units together in their random assemblies, like pieces of a child's construction kit.

The units themselves will be infinitely adaptable. Windows, doors and even walls will be arranged according to individual taste to derive the maximum benefit from their particular setting. One unit might be large enough to provide a complete bed-sitting room for one student. On the other hand it might be used in conjunction with one or more others to form larger family rooms. If your mother-in-law comes to stay, if the children become difficult or if you simply feel in need of a little more space in the home, the answer will be perfectly straightforward. There will be no need for a second mortgage while you try to build an extension, no need either to move house. You will simply buy another basic unit, add it to the others and adapt it to whatever purpose you wish.

As the family grows, your home can grow with it, assuming that the planning authorities permit you to sprawl your cubist's ideal across the countryside. On a larger scale, schools, shops, hotels and offices could be constructed from these multi-purpose units. When all is said and done, though, they will still bear a startling resemblance to 'little boxes on the hillside' even if they are not 'made out of ticky-tacky'. In my view the most satisfactory development in housing policy would be a recrudescence of urban architecture.

Housing policy is notoriously slow to develop. One third of the houses in England and Wales are over fifty years old, and there is little reason why this should not be true in the twenty-first century. Now I am very fortunate living where I do. St Peter's Square,, in London, was built in 1830, 200 years before the time I am writing about. I have been living there for forty years and people have always wanted to live here. It does not matter whether the social level has gone up or down, it has always been a popular place to live. The houses are all terraced with two, three or four upper storeys and a basement. This would be an ideal arrangement for civilized human living in the year 2030. I was born in a house that was six storeys high and I think that is just about the limit. We have revolted against the practice of piling people on top of each other in tower blocks and I think that the long-standing adulation for semi-detachment will also fade. Single-storey houses, on the other hand, are uneconomic and do not make for contented living, because your house is scattered around. Suppose you happen to be at one end sitting on the patio when the milkman rings at the front door. You have to go trapesing through from the patio to the lounge, opening and closing doors, stepping around the furniture and avoiding the children and their toys littered across the floor. Whereas if you were upstairs when you heard him, you could come straight down the stairs and let him in. There would be less fuss and your home would be a more ordered, less disturbed little unit. We need to persuade the urban planners that the bulldozer is not necessarily the answer to slum clearance. After all, slums are only houses in which slum-dwellers live. There is a movement now to 'gentrify' old houses in working-class districts and perfectly simple houses have been turned into very desirable properties, in the jargon of the estate agents. This practice will be as common in the 2030s as it is in the 1980s. It is all a question of prosperity.

This is not to say that if you are poor you will be living in a mud hut, but I anticipate that housing will be very much simpler in the future than it is today. At present, we claim to be so wealthy that we build houses with costly underfloor central heating and other luxurious features, which the inhabitants are often unable to afford to run. In the future people will be offered greater choice on how much they want to spend on their home. If, as I expect, we are all going to have a lot of free time – in other words, time when we are not working for money – a lot of people will choose to take the opportunity of creating their own homes.

When our descendants go along to the housing officer to ask for houses he will not tell them to join the bottom of the waiting list, he will say, 'Right you are chum, go round to the do-it-yourself shop and get on with it,' and off they will go. The council will provide the basic shell and the essential services and the householder will go staggering home from the D-I-Y shop with a bath tied to the roof of the car and a lavatory in the back seat.

Many people who first foretold these ideas have now discarded them, but I feel confident that with the great improvement in building materials which we will see in the next fifty years the problems that currently deter the would-be home-builder will have disappeared. Creating homes in this way would give people endless choice on the interior design. They would be able to select their sitting-room kit, their kitchen kit and their bathroom kit, all in prefabricated form. How elaborate these are and whether they have stained glass windows painted by a privately commissioned artist or just plastic sheeting to keep out the draught would be a matter of taste and affluence.

Although the choice of shape and dimension would be very much the decision of the individual concerned, and I have no wish to influence his frame of mind, I think that the current vogue for open-plan designs will disappear. I should like to think that people will start to realize the virtues of living in a house with comparatively small rooms that provide privacy, quiet and warmth.

By means of the do-it-yourself home building system, householders will be able to construct one room after the other and decide on their functions later. They will be able to make use of mobile partition walls, like those in Japanese homes, to change the dimensions of different rooms to suit different occasions. If the children are playing with friends in the living room, their parents will only need to push back the wall to make room for them all and then secrete themselves in what remains of the dining-room to enjoy undisturbed peace and quiet. Underlying this whole principle of house building is the adoption of new building materials which will move the construction industry at a domestic level from the province of the expert to within the resources and capabilities of the average handyman.

Invest in cardboard and plastic

The reason why the Japanese have been able to move the partition-walls inside their homes with such facility is that they have made the walls out of paper. Of course this is a wise precaution if you happen to live in a country which is liable to undergo an earthquake at the drop of a hat. But since there is considerably less risk of violent subterranean upheavals in Iowa or Wapping we might afford ourselves the luxury of walls slightly more robust than those made from the sort of material through which performing dogs leap in circus arenas.

A very satisfactory wall could be made from a honeycomb of stiffened paper sandwiched between two sheets of thick cardboard. This would be light and easy to move about, yet it would be strong enough to support all the normal wall fittings and even a lightweight door made from the same material. Paints and self-adhesive wallpapers would adhere easily to its surface. It could be treated with fire-proofing material to minimize the risk of domestic conflagrations and if the cavities between the honeycomb ridges were filled with insulating 'wool', walls made from this material would be very effective in trapping heat inside individual rooms, helping to reduce the overall consumption of domestic heating fuels. These walls would be simple and cheap to construct. The average do-it-yourself enthusiast would be able to buy a standard wall, which he would then be able to cut and shape to any design he had in mind. If after a few months he should decide to rearrange the design, he will merely dismantle the wall, take it back to the D-I-Y shop to be recycled and come home with a brand new set to start all over again. There will be no clouds of choking dust, no incessant hammering out of bricks and no piles of rubble to block the hallway or litter the garden. Home improvement will be looked upon with the hopeful anticipation of rewarding work and rapid completion, very different from the sense of fear and loathing that grips most of us today at the prospect of major alterations in the house.

Problems experienced with plastics at present will almost certainly be overcome by the year 2030, which will extensively broaden the range of their uses in buildings of the future. Plastic roofing has already reached the garden shed and the car-port; in fifty years it will have ascended to the top of the house as well. In the past we had slates, then we used asbestos and clay tiles – all of which are expensive. Plastic is the obvious answer. It is cheap, light, durable and waterproof. Plastic pipes will have completely replaced copper pipes in plumbing fifty years from now; they almost have already. We might even have to think up a new word for 'plumbing' which is really derived from the Latin word for lead, and lead has not been used in plumbing for a long time.

When our grandchildren are fitting-out their bathrooms they will not have to resort to calling in experts to connect their baths and washbasins to the water supply, they will have simply couplings that lock the plastic pipes together without the use of heat or sealers.

I remember an advertisement I once used in a lecture I was giving. It was proclaiming the powers of a well-known Irish stout, and underneath was a picture of a workman carrying an enormous steel girder on his head. The joke was that no one could possibly carry anything that size on his head, no matter how much he drank. In the lecture room, I had a large drainage pipe, about twenty centimetres in diameter, which the laboratory assistant and I easily lifted between us. If this great pipe had been made of cast iron, enormous lifting machinery would have been required to pick it up, but it was made of plastic. So in the future I foresee a similar use being applied to roof structures. We will have plastic 'wonder-rafters', and 'wonder-beams' made from plastic reinforced with carbon-fibre, which is also strong and light. Currently, we use wood to construct our roof frames. Wood is a natural 'plastic', a polymer made from lots of little glucose molecules, which is what cellulose is. Wood is a very elegant, useful building material, but unfortunately we have to wait thirty or forty years for it to grow. The answer is to do the job just as well with plastic, making use of the oil we retain for the petro-chemical industry, provided we do not burn it all up.

At present, it is fashionable to want our plastics to be biodegradable. If you are sufficiently anti-social to strip the plastic wrapper from your cigarette packet and nonchalantly toss it away, you may be comforted by the thought that in the end it will wither away and disappear like a dead leaf. Unlike contemporary houses, the houses of 2030 will be 'non-biodegradable' that is, once built they will not change shape or be subject to the disintegrating influences of rain and wind. Today our central heating dries out the floor-boards and the timber window frames causing them to shrink and let out the warm air we have produced at such expense. However, our new construction materials are not going to shrink. When you close a front door in fifty years time the effect will be the same as shutting a car door today. When you bang one door the opposite door pops open, and if it does not pop open, it pops your ears. The nylon socks we wear do not shrink when we wash them because they are made out of petroleum, and our houses of the future will not shrink for the same reason.

Keep the home fires burning

Experts are constantly telling us that we are grossly extravagant with domestic energy. Central heating in the winter and air conditioning in summer frequently maintain temperatures higher or lower than are really necessary for comfortable living. But with energy shortages threatening to curb our wasteful excesses, many people are looking to the sun, the giver of life, to be the giver of domestic heating in the future. They look forward to the time when no house will be complete without its solar panel in the roof. Activated by the sun's rays this miraculous unit will heat our bath water and provide auxiliary heat to the central heating system. Some wilder flights of imagination see it performing even more impressive tasks. So, although solar power is dealt with at greater length later, it is worth considering its domestic application here.

One of the very first fully solar-heated houses, suitably called Solar One, was built at the University of Delaware in the USA, to test the feasibility of using sunshine to generate both heat and electricity. The house is a single-storey structure with an unusual high sloping roof, set at an angle of 45° to achieve the maximum exposure to the sun. Solar cells in the roof panels are used to generate electricity which is either used immediately or is stored in batteries. Heat is 'stored' in salts which melt at a relatively low temperature, 20°–50°C. In the evening when the temperature in the house drops, air is circulated through these salts to be warmed by the heat contained in them and then re-directed through the house.

This experiment has in fact proved to be very successful and solar-heated houses can now be found in several parts of the United States and Europe. Four-fifths of the energy requirements of Solar One can be met by solar power alone and the rest is made up by conventional heating and lighting appliances. However, Delaware lies south of the fortieth parallel and most of we wasteful energy consumers in the developed countries of western Europe and the northern parts of the USA live north of this line. Unfortunately this change in latitude makes a considerable difference in the efficiency of solar heating. For the angle of elevation over 45° north severely limits the power of the sun during the winter months, the coldest part of the year, when most of the heat and light are required. With existing solar panels, alternative heating and power systems have to be installed in these latitudes if the inhabitants are to live at all comfortably, and this reduces the solar power system to the level of an expensive optional luxury. This does not mean that solar energy will never be used in these parts of the world, but it will have to become a good deal more efficient and economical to install. Fifty years may not be long enough for the technology to develop to meet these demands.

The most practical answer to the rising cost of home heating must be insulation. We have only recently started to promote insulation through advertising and the offer of local authority grants, although there are materials in existence that would be capable of keeping a cup of coffee warm for 2000 years. Our current building regulations in Britain and the USA tolerate a high rate of heat loss from most houses built to conform with the minimum insulation requirements. At least 20% disappears straight through the roof; the same amount is lost through the walls; 15% escapes through the floor and the rest goes out through the doors and windows.

By taking a few simple precautions this loss could be halved, and, in the future, measures like these will be commonplace in every building. We will be laying a minimum of 75 mm of fibreglass matting in the loft to reduce heat loss through the roof to one-fifth of its previous level. Cavity walls insulated with mineral wool will lose only one-third of the amount of heat that goes out through a solid brick wall and one-sixth of the amount lost through the windows.

In the case of windows themselves, double glazing has been the only measure taken to improve their heat loss up until now, but this is often expensive, especially in old houses. In the future two things will happen to windows. Firstly they will become a lot smaller. Picture windows have not been a great success and people on the whole do not like them. Secondly, window glass will be different. I can conceive of a great revolution in glass in the next fifty years. We may see glass as thick as that in Liverpool Roman Catholic Cathedral in our future homes. That would be one solution to the loss of heat by conduction. On the other hand, hollow panes of glass might be developed. A 'float sandwich glass' process would do away with expensive double glazing, because it would be double glazed already.

Great advances have been made already in the glass used to make motor car windscreens compared with the glass they put into the old bangers that were around when I was a boy. New domestic glass will be satisfactorily transparent, but another big improvement I see for it will be that it will never get dirty. It will be self-cleaning – the rain will just hit it and bounce off taking the dirt with it.

The window-frames themselves will be made from non-expanding plastic or metal. The wooden frames most of us use at present have rather poor tolerances in engineering terms and their dimensional fit cannot be very good. Improved frames would cut the heat loss there. They will not expand when it is wet and shrink when it is dry, nor will they require painting every three or four years. However, since many houses will still have wooden window-frames in 2030, the cheapest contribution to keeping heat in the house will not be double glazing but good thick curtains.

Once we have mastered the insulation of our homes the problems of heating them will be minimized immediately. There are flats in Sweden that can be kept warm by little more than the heat given off from the electric lights, the cooker and the gas pilot lights, even when the temperature outside is well below zero. The only difference between these flats and any others is that they are well insulated.

We can contribute to heating our own homes just by living in them. We all produce a great deal of heat because we are biochemical engines. As we sit munching a biscuit, which is primarily carbohydrate, we are breaking it down and this produces heat. Just as the car engine ticking away outside gives off carbon dioxide and water through its exhaust pipe, so we give off carbon dioxide through ours – that is, through our noses and mouths. Like the motor car with its radiator we give off heat into the atmosphere.

So in the future we will be heating our houses just by walking around and living in them, leaving a trail of warm air wherever we go. Efficient ventilation systems similar to the ones currently used in Sweden will ensure that condensation is minimized and that the air remains fresh.

The electric lights we use will also be giving off a considerable amount of heat. At present a mere 5% of the power that goes into a light bulb is used to produce light. The remaining 95% is lost as heat. Even if great improvements can be made in the design and construction of light bulbs, it is doubtful if they will have made any appreciable difference in fifty years. It has taken all of that to raise the level of efficiency from 1% to its present level. When it gets very cold, though, we will need some form of supplementary heating, and much of this could be obtained from existing domestic appliances.

At present we have a great many refrigerating machines. These work on the basis of the Carnot cycle which was developed by Nicholas Carnot back in the 1800s and gave rise to fish fingers and space rockets, as well as everyday fridges. The operating process is very straightforward. You merely pump heat out of the refrigerator to keep it cool, and pump more heat more efficiently out of the deep freeze to keep it cool. The heat is pumped out by the evaporation of a refrigerant in one place and its condensation in another, but nothing is done with the heat pumped away.

However, by means of a small modification which is already available, it is possible to pump heat out of your freezer and use it efficiently in the home. This modification greatly impoves the effective use of the electricity needed to drive the freezer by a factor of about five.

Heat pumps will become increasingly common in their own right. When employed thus, they will be using the elements – earth, air and water – as their source of heat, not merely the heat extracted from the Sunday joint or the Christmas turkey when it is put into the fridge. The fact that these elemental heat sources are not very hot is not important. What is important is that there is a lot of them.

In a heat pump the heat, whatever temperature it may be at, is used to evaporate an appropriate heat-transfer medium, a type of refrigerant. The evaporator will either be buried in the earth, immersed in water or exposed to the open air, whichever source of heat is selected as most suitable. The evaporated refrigerant is then drawn as vapour into a compressor, where it is compressed, and passed into a condenser, where it is converted back into liquid, and, in condensing, gives off heat. The cycle is completed by the refrigerant passing back through a valve into the evaporator, where it is again caused to evaporate by the heat of the air, water or earth at the lower pressure of the partial vacuum there.

If I am right in my ideas for the future, the prudent householder of 2030 will ensure that he has at least one open fireplace in his house and a pile of dry firewood in the garden shed to burn in the event of a particularly sharp cold spell or a power failure. For although we may have sophisticated heating systems and other highly complex pieces of domestic machinery we will have learnt by then that technology is fallible and that machines can and do let us down, often at those times when we are most in need of their services. On these occasions we will be resorting to the important safeguards of the future which I have initialled GASH (Guards Against Social Hitches). These simple, practical devices will ensure that we are able to carry on our normal lives, little inconvenienced by the hiccoughs in everyday life in the future.

So if our householder of 2030 wants to be really efficient

he will also install a small garbage compressor in his kitchen. This simple mechanical device could make an enormous difference to garbage disposal in fifty years time. It would probably look like a large pedal bin with a lifting lid and a handle-cum-pedal lever on the outside. Instead of throwing the cornflakes packet into the waste-bin, thus virtually filling it, and instead of waiting for the Boy Scouts to come and take away the piles of old newspapers, he will put them into the garbage compressor and squeeze them into little combustible brickettes. After all, paper is only wood and these little brickettes will make a very useful addition to the firewood when the north wind blows and fuel stocks run low. Any amount of household refuse could be disposed of by this means, providing that it will burn. So apart from making a fortune for the inventor, this perfectly simply machine could save thousands of pounds that might otherwise have to be spent disposing of the garbage by conventional means.

Another simple machine that will very probably become a standard domestic article in the future will be a small standby generator which would be run off bottled gas or cans of petrol or paraffin stored underneath the stairs. Like the garbage compressor, this will be a very convenient form of self-help when the electricity fails or the men who operate the power stations decide to go on strike. Under normal circumstances our lives would grind to a standstill during a power failure. There would be no light and no heat. Even if the central heating and cooking worked off gas we would still freeze because the pump which brings the gas into the house and the circulating water pump for the central heating also work off electricity. With this little generator, however, it would be possible to keep storage batteries charged at a level at which they could run emergency lighting and provide the small amount of current needed to operate the gas pump. If the power cut lasted to the point when we ran out of fuel, it would be possible to drive the little generator by bicycle power. This would be adequate to meet domestic needs. There is already available a pedal-powered generator producing sufficient electricity to run a television set. The greatest domestic catastrophe we can suffer is the loss of the television. So if the lights go out in the middle of the favourite serial fifty years from now, the remedy will be at hand. We will wheel the generator and the bicycle unit into the sitting-room, connect the television lead and the bicycle drive-belt to the generator and start pedalling. The family will be delighted and we can get off and pant during the commercials.

Kitchen wear

Though the traditional picture of the mother who spends a good deal of her time in the kitchen is rapidly disappearing, rightly or wrongly there are many women who still spend many hours working there. However, optimists tell us that their daughters and grand-daughters will be completely freed from this necessity. Kitchens, they claim, will be transformed. Hours of bending uncomfortably over a badly designed sink or slaving in a lather of perspiration (or 'glow' as the adage quaintly phrases it) over a hot stove will be swept away by micro-technology. The glossy kitchens of the colour supplements, with their rows of dials and switches, and their elaborate gadgets for beating eggs or taking the tops off milk bottles will be available to every one of us in the future.

Even the dishwasher, invented way back in the 1880s, will be superseded by a new sophisticated ultrasonic model. Messy overflow pipes and strong detergents will become distant memories of a past era when power and water were cheap and plentiful. In the kitchen of 2030 the lady of the house will stack her plates in a rack and push them into her dishwasher, just as her grandmother used to. She will even use a little water and a small quantity of very mild detergent. However, when she switches on the machine her kitchen will not be filled with the monotonous rumbling of the agitator or the disconcerting swish of the present-day water jets; she will turn on sound waves instead. After a few minutes of ultrasonic cleaning the cycle will be completed and the door will pop open revealing pristine crockery, sparklingly clean and dry.

Similar predictions indicate that sophisticated machinery will transform the art of cooking as well. The microwave oven, which has been described by Margaret Costa as 'the first absolutely new method of cooking since prehistoric man invented the making of fire', will seemingly sweep through the kitchens of the next fifty years. These glass-fronted boxes that miraculously transform a deep-frozen pie and beans into a steaming meal in seconds will be reflecting the happy smiles of housewives everywhere, as they close the door on the prepacked, precooked and probably predigested evening meal, confident in the knowledge that what took their mothers forty minutes to prepare will be theirs in as many seconds. Labour-saving, clean and sophisticated, the technology of the microwave oven of fifty years' time will, we are assured, have overcome the few remaining problems which limit its general use at present.

Another forecast sees the introduction of induction heating into cooking. Place your hand on the ceramic surface of the hob and it feels perfectly cool. Put a steel or an iron pan on to the seemingly cool hob and it becomes hot. The heat is generated by the resistance in the pan to an electric current which passes into it from a magnetic field. The magnetic field is itself produced by an alternating current passing through a coil in the ceramic substance. This method of cooking requires less energy than conventional stoves because the heat passes directly into the vessel being used to cook the food. If the milk boils over or you accidentally put your hand on to the cooking surface, there is no smell and no need for alarm.

If these dream kitchens do in fact become commonplace, they will be labour-saving, hygienic and fitted with an array of equipment that will make them a far cry from the relatively crude kitchens of the 1980s. But will they be places where we will be able to relax, or even wish to relax? I think not.

In fifty years' time we are going to be relatively richer and relatively more sophisticated. After all, the whole purpose of our efforts is to improve the quality of life. Labour-saving devices have one primary aim, to save labour and by so doing create leisure. Yet, having created leisure, we have to do something with it. I suspect that in the home of the future, manual skills and craft activities will take on increasing importance in our leisure time. Of these craft entertainments, cooking will be one of the most satisfying and popular. Today we have precious few manual skills left, but those that have survived the onset of automation and technology are jealously guarded. There are many middle-aged men who can produce wonderful omelettes. They flip and flap away with wooden spatulas and forks for entertainment. Just as we enjoy the challenge of catching trout by fly-fishing, we may come to relish that cooking which causes the greatest amount of difficulty. After all, it is much easier to catch trout with nets than with the delicate paraphernalia we use today, and the same is true with some continental cooking in which you make things artifically difficult by trying to do a miscellany of different operations on the food before you eventually sit down to eat it. Under these conditions, the last things we are going to want are labour-saving devices. If indeed these clever contraptions are available to us, I anticipate that we will in fact choose the contrary direction, eschewing the quick and efficient in favour of the complicated and demanding.

We may well go back to cooking in kitchens with rows of copper pots of all shapes and sizes hanging on the walls. Technologically these may be highly inefficient, but from a craft point of view they will be fun to use. This could lead to a situation in which there are fewer sophisticated gadgets and tools in future kitchens than are in use today. There are many domestic appliances available to us at present that reduce the effort of daily chores and no doubt perform these tasks more efficiently than the tools that preceded them. Nevertheless many of us still prefer to use the old-fashioned, general-purpose tools, and we will continue to follow this practice in the future.

Earlier I mentioned dishwashers. Owning and using a

dishwasher may mark the limit of technological complexity. Leaving aside the fact that the machine is vulnerable to failure, both mechanical and human, if the power is switched off, washing the dishes can be a wonderful social exercise. If you are having a party with real friends, you do not break it up by saying, 'Now do come into the lounge, won't you.' You all go as one man to the kitchen carrying out the dishes, talking, laughing and finishing off the last of the wine. The happiest moment of the party may be when you are all gathered together round the sink, one washing, another drying, and the rest putting things away.

Sinks themselves will be better designed in the future than they are today. They will be installed in a way that will enable the height to be adjusted to a comfortable working level for anyone. The actual sink materials will alter as well. When I was young sinks were made of ceramic. Today they are made from stainless steel, but both materials are a bit of a nuisance to clean. So in the future some improved plastic will be developed to prevent dirty marks forming and obviate the need for these strong abrasive detergents.

There are some people who argue that dishwashing itself will have disappeared in fifty years' time. They tell us that we will be eating off disposable plates and drinking out of disposable cups. Present experience shows that there is a revolt against disposables in general and disposable crockery in particular, and this revolt is gaining momentum as we become aware of the extravagance of disposability.

Crockery is pleasant to eat off. Its major disadvantage is that it breaks if you drop it. Nevertheless, during the next fifty years technological advances in ceramics may well produce a china-nylon combination, rather like the windscreen of your car, which is marvellously strong. These developments may not only take place in the fields of technology and science, but will be seen in our homes. In ceramics, as in all aspects of life, there will be a greater awareness and appreciation of aesthetics. It is difficult to make elegant disposable articles; furthermore, if they were elegant they would not be disposed of. It would be wonderful to wake up in fifty years and find ourselves in a new

Florentine era, when the everyday objects of life were not only strong but beautiful.

When my son started a hotel business ten years ago, my wife said, 'You need the most beautiful crockery,' and she gave him some. But within a month the customers had chipped or broken it all. For the catering industry, crockery may be elegant but it must be strong. In fifty years' time we shall see something similar in our homes, too.

There will be equally elegant cutlery to match this decorative tableware. Dining at a great banquet with all its beautifully polished cutlery, or eating with Georgian silver in a stately home is an agreeable experience. In the future, cutlery like this will be the cutlery of fashion, the lubrication of social well-being, as it can be today. If the cutlers of Sheffield know what is good for them, they will start designing this beautiful, strong, cheap cutlery now.

One significant development that I do envisage is the development of a type of multipurpose, utilitarian cutlery tool, that would offer the great advantage of being able to eat anything with one hand. Great developments have taken place in the design of cutlery for the handicapped, the crippled and the mentally deficient. There are now handles that actually fit your hand. If you suffer from arthritis you can now feed yourself perfectly easily with a custom-made tool that both cuts the food and conveys it to your mouth. Something along these lines could be very useful in the future for people who wish to use one hand to shave or read their letters in the morning, while eating their breakfast using the other. This dual-purpose tool would probably look like a cheese knife, with a fork on one side and a knife on the other.

The growing awareness of aesthetics may help in changing the kitchen from an antiseptic clinical workshop in which technological marvels are performed, into a beautiful environment for living. Many people take a pride in their kitchens and adore beautiful polished furniture, and as part of our improved quality of life we may well see more polished walnut and mahogany in our kitchens. They may be wood surrogates, because real wood will become rather rare and unusual, but improvements in plastics and artifical fibres will make them look like wood. For those who can afford the luxury of wood there will be a protective table-cloth made from some impervious material to protect the delicate surface. This cloth will not require laundering or ironing – just wiping clean, after which it can be folded away into a drawer.

The other principal function of the kitchen will be to act as a storehouse. Fridges and deep-freezers will be cooling our food and circulating the warm air through their heat pumps into the house. When they cease to operate through strikes or breakdown, we shall neither panic nor starve. We shall simply go to the larder in which we will have prudently stored all the supplies necessary to protect ourselves against the periodic wrinkles in the social fabric. Like the fireplace, the generator and the garbage compressor, the larder will be an essential GASH fifty years from now.

Bathtime

If the present-day kitchen is a place of work and anguish, the bathroom at least is a retreat of luxurious self-indulgence. There, locked away from the television and the kids, we can immerse our tired bodies in the sensual delights of warm water and fragrant cleansing agents. Louis XIV may have taken only three baths in his life and fastidious Queen Elizabeth I only one a month, but civilization has developed since then. We have come to appreciate the soothing feeling of wallowing up to our necks in dirty water, singing arias from *La Bohème* and idly sending plastic ducks drifting down to our toes. Can such enjoyment be improved upon? Future-watchers tell us it can.

According to some predictions, washing ourselves in a bath will seem as primitive in fifty years' time as cleaning knives in a knife-cleaning machine; those who currently advise taking a shower after a bath no doubt share these sentiments. There is one idea that we will be climbing into a man-sized bubble in which a cocoon of mist will cleanse and relax us before we are blown dry by jets of warm air. Another envisages us receiving more than a simple cleaning at bathtime, we will be given a full massage at the same time.

A machine capable of performing this exhilarating task has in fact been developed in Japan. Shaped like a large egg, it cleans the whole body, except for the head which sticks out of the top like the knob on the lid of a kettle. Sitting inside this 'bath' of the future will be rather like sitting inside a giant human dishwasher. Once installed, we will be given a warm shower for two minutes. This will be followed by the egg filling with warm water. Ultra-sonic waves will be passed through the water causing it to bubble up around us and clean the dirt from our bodies, just as the dishes might be cleaned down in the kitchen. After this there will be a second wash in swirling warm water while the ultrasonic waves continue their silent work. At the same time small rubber balls will be released into the water to pummel our muscles and skin, relaxing them with a mechanical massage. After a final shower to clean off the last of the grime, warm air will circulate inside the egg to dry us, and fifteen minutes after closing the lid round our necks we will open it again to emerge clean and relaxed, ready to face the world. There will be no puddles or damp bath-mats and no steam to peel the paper from the walls.

There are many people, though, to whom the idea of being bombarded by unseen rubber projectiles inside an electronically controlled egg is abhorrent. Similarly, there are others who enjoy longer or shorter baths than the rigid fifteen minutes allowed by this machine. Perhaps the luxury of lying and soaking in warm water will become more widely recognized. The mechanical body-washer may, therefore, come to be regarded as an amusing novelty for the wealthy. I think most of us will be happily climbing into our baths in fifty years. (The really socially conscious among us will probably only be taking showers, because they are far more economical than baths. Nearly 10% of the total energy consumed in a house is used to heat water).

This does not mean that bathroom technology will stand still. A major advance that will have taken place by 2030 will be the development of a ringless bath, which, like the ringless sink, will be self-cleaning. Dirty 'high-water' marks will not be able to form and we shall not be obliged to scrub with abrasive cleaners to remove our grimy deposits. Another avenue of engineering inventiveness that has been sadly neglected so far is the non-drip, precision bathtap. Doctors can turn off taps with the elbows and on trains it is possible to turn on the tap with your feet, but domestic taps remain inefficient. They drip, and when they do, they tend to stain the bath. Such domestic troubles will have been overcome in fifty years' time.

As far as washing is concerned there is a great advantage in using soap because it is made out of renewable resources, fat and soda, unlike detergents that are made from fossil fuels like petroleum. Still, some ingenious detergent manufacturer may invent a detergent applicator, which would be more convenient than a cake of soap. In order to work, this applicator would need to have a liphophilic end, that is, one that picks up the oily dirt on our bodies, and a hydrophilic end, which picks up the water, because what soap does is to make the oil and the dirt mix with the water in the bath. Like the plastic lemon which contains lemon juice, I expect that this applicator would initially resemble a cake of soap and be pleasant to hold. In the end the clever manufacturer will probably develop the existing bath glove to work like a giant Brillo pad. This might contain a flexible cleansing cartridge that could be zipped in and replaced once it was exhausted.

Once out of the bath we shall dry ourselves, as we have always done, with cotton towels. Hot air drying is costly to install, wasteful of energy and leaves you not feeling dry; whereas cotton towels feel pleasant, rub off the dirt we have not fully removed in washing, and efficiently absorb moisture. The fact that nylon is much easier to wash and dry than cotton shows that it absorbs dirt and moisture less effectively. No, cotton towels will certainly last fifty years. They may even become the sign of a well-to-do household.

Instead of subjecting one's guests to a blast of hot air they will be offered cotton towels on which to dry their hands.

The smallest room in the house

The disposal of human excrement has been a matter of concern and complaint since the dawn of history. Ancient civilizations in Crete and the Indus valley succeeded in providing adequate drainage systems in cities like Knossos and Mohenjodaro 4000 years ago, which appeared to have dealt with the problem. The Romans, too, made use of communal lavatories which relieved their larger settlements of the hazards and smells that were a part of everyday life in the Middle Ages. The situation became so intolerable in Paris towards the end of the fourteenth century that a royal decree banning the throwing of excrement out of windows was issued in 1372 and again in 1395.

One of the earliest pioneers of the water-closet was the godson of Queen Elizabeth I, Sir John Harrington. He published a treatise in 1597 describing the function of a W.C. he had heard about in Italy. His godmother was delighted and immediately installed one in Richmond Palace. As a woman whose preoccupation with personal hygiene was a source of constant amazement to her courtiers and perpetual misery to herself, she was finally allowed to spend her declining years free from the offence of her own ordure.

However, the innovation never caught on. The entire Palace of Versailles, the grandest building in Europe, was constructed in the seventeenth century without the inclusion of a single lavatory. For the most part, people seemed oblivious to this dearth of 'conveniences'. They either went to pay their calls of nature in the privy in the garden, or more often than not in a chamber pot behind a screen. The rise in the use of W.C.s only came about as a result of municipal drainage systems that evolved during the nineteenth century. Ironically, like so many developments of the

Industrial Revolution, they are now posing a serious threat to our future.

The widespread and in many cases profligate use of water to transport sewage has placed our future water resources in jeopardy and will continue to do so unless we rapidly mend our ways. In fifty years' time we are bound to be far more water conscious. The so-called droughts in western Europe during the mid 1970s gave us a very unpleasant lesson on what it means to live with a restricted water supply. At present, we still run taps unnecessarily, take baths when we could take showers and waste millions of litres transporting and processing sewage which, as the Chinese have shown all along, could make a valuable contribution to fertilizing our crops.

The Swedes, perhaps the most environmentally conscious nation on earth, have developed an ingenious system of overcoming this waste of water. Instead of carrying away domestic sewage and kitchen waste in a stream of water, they use an invention to incinerate them on the premises.

All the waste water is collected into a large storage tank, with the exception of the W.C. water which is kept separate. This, and the scum filtered from the other household water, is incinerated in a boiler until it is all reduced to ash, which can be disposed of easily in a dustbin. The rest of the water is distilled and recycled through the heating system and the W.C. In order to economize on fuel, the whole incineration plant can be integrated into the existing heating system, and a further advantage to the householder is the fact that it requires no exterior plumbing, avoiding the need for drains or septic tanks. However, the plant still relies on the use of expensive fuels. If these are in short supply in the future such a unit could rapidly become obsolete before it ever came into general use.

If we want to reduce the amount of water used in our W.C.s and also decide to put sewage to some useful purpose, the answer is surely to install lavatories like the ones currently used in aeroplanes, where there is an obvious need to economize on water.

When there was a drought in California a few years ago,

the authorities plastered the state with a slogan that would be very apt in encouraging the use of the sort of W.C. I have in mind. The slogan ran, 'If it's yellow, let it mellow. If it's brown, flush it down.' There are lavatories at present that partially do this by means of a controlled flush, but in the future the water will come out under pressure in the form of a jet, which will be much more efficient.

Not long ago, W.C.s were operated by pulling a chain. This was noisy and tended to disturb the household. The lavatories of the future, I am glad to prophesy, will be more discreet and will economize on water as well.

And so to bed

The design of beds is very much a matter of fashion and fashion is notoriously difficult to predict. There are some who believe that we will be spending our future nights in 'pleasure domes'. These sophisticated offspring of technological imaginations will apparently relieve us of all our current night-time problems. They will contain automatic controls to play gentle music, or soporific lullabies to send us to sleep at night, and in the morning they will rouse us just as gently by automatically drawing the curtains, making our tea and switching on more lively rhythms to prepare us for the day.

Less fanciful predictions herald the introduction of hover-beds. Tucked up in these we will spend the night 'floating' on air. A cushion of warm air heated to the preset temperature will circulate underneath us all night, providing unimagined comfort and presumably unmatched sleep.

While inventions like this are no doubt highly ingenious, I foresee a general increase in the unreliability of high technology, especially when applied to the domestic scene. The people who install these clever, costly pieces of furniture are going to feel very uncomfortable when the airpump packs up or the waterbed springs a major leak. In these instances an ordinary bed could be a valuable GASH. The one great advantage of the ordinary bed is that it does not require a great deal of maintenance, nor does it require an outside source of energy to operate it. Years ago, my mother used to say, 'Magnus, you ought to turn your mattress every day because you will be more comfortable,' but that was maternal propaganda. The bed was just as comfortable whether I turned the mattress or not.

The present bed has been with us for a good long time and broadly speaking it has remained much the same. There are divan beds only a few centimetres off the floor and there are high, old-fashioned brass beds that are becoming popular once again. The bed that I sleep in is a four-poster that once belonged to the only Archbishop of Canterbury who ever hunted his own pack of fox-hounds. This was Archbishop Juxon, who shrove Charles I on the scaffold. There are modern springs and a mattress in it now, but otherwise it is exactly the same as it was when he slept in it three hundred years ago. Like all other beds it is merely a comfortable surface on which to sleep. Though I dare say in the future the springs will be ingenious plastic or foam-rubber affairs, replacing the old-fashioned and sometimes noisy metal ones.

Bedrooms themselves, particularly British ones, will be warmer than they are at present; improved insulation will make them so. We shall not therefore have to crush ourselves as flat as the Jack of Spades with a pile of blankets in the winter, nor will we be obliged to purchase costly duvets as an alternative. Our only covering will probably be a snug nylon sheet. We will not be poaching ourselves on top of electric blankets either. In fact, the electric blanket is merely the technological successor to the rubber hot-water bottle, which was itself the successor to the stone hot-water bottle, which was the successor to the warming pan, which was inconvenient, smelly and a fire hazard. This only goes to show the inefficiency of domestic heating in history and the urgent need for its improvement in the future.

Just as beds will remain relatively unchanged in fifty years' time, so will the overall design of the bedroom. As in other parts of the house, the walls may be covered in light-conductive paint or paper, to help reduce the amount of domestic lighting. It will no doubt be possible to 'operate' the wardrobe from the warmth and comfort of your bed in the morning. Automatically opening doors and automatically sliding racks of clothes will display the contents of the wardrobe for our selection. The chosen garments will be lifted from the rack by a mechanical hand and placed on a conveniently positioned stand. The same machinery will no doubt be capable of performing the reverse operation for those too lazy or too incapable of putting away their clothes after they have been worn.

Dressing-table mirrors may well be lit from behind to produce light on the face that comes directly out of the glass. It might be possible to examine one's make-up in an artificially created daylight before leaving the dressing-table. Indeed there are some designers who foresee the development of artificial 'sunlight' throughout our buildings before the end of the century.

The walls may be decorated with giant oil-slides similar to those projected on to some cinema screens. These would create an infinity of patterns and images that would be controlled by a thermostat. On going to bed and telling the lights to dim (see page 39), the room might be bathed in the light and colour of these giant psychedelic images that would slowly cool and fade as they assisted us to fall asleep, mesmerized by their almost hypnotic power.

But as I have said elsewhere, although designs and artefacts like these may be available to the gadget-conscious and gimmick-oriented in the future, it is my strong belief that we will continue to sleep and dress in bedrooms that look very much like those in which we are quite happy to live at present.

Domestic service

Fully automated home-helps which perform the mundane tasks of daily housework are already under development in research laboratories. Professor Meredith Thring, one of the leading designers of the machines that will make life in the future very different from what it is today, has already built prototypes of home robots which might be doing the domestic chores in the twenty-first century.

These taciturn, mechanical 'char-ladies' will be capable of making beds, vacuum-cleaning, scrubbing floors, cleaning ovens, clearing tables and even plugging themselves into the mains to recharge their own batteries.

Extendible arms equipped with spring-loaded hands will be able to grasp safely most household objects. Small television cameras or sonar devices will enable the robots to move about without bumping into the furniture, running themselves against the walls or falling downstairs. Each robot will be programmed to meet the individual needs of the particular household to which it is on hire. It will store in its memory details of the arrangement of the furniture in every room, the position of the electric sockets and the location of all the objects on the shelves and tables. If any of these is moved, the robot will be able to return it to its correct position. A simple reprogramming is all that will be required when the furniture is rearranged or the Christmas tree set up once a year.

The robots will be fitted with sprung spokes so that they can climb normal stairs to clean the bedrooms and make the beds. They will be able to carry dustbins down to the street, and move heavy pieces of furniture. Their mechanical hands will be equally dexterous in unscrewing obstinate lids and bottle-tops or lifting priceless Meissen figurines with confident delicacy. In the end, our only concern with housework will be to set the controls for the mechanical minion and then sit down with a cup of coffee while the robot gets on with the work with far greater efficiency and enthusiasm than some of us are ever able to muster. Our lives will be carefree and, above all, housework-free.

Nevertheless I am certain that in the year 2030 many of us will be happy to continue to do our housework once it ceases to become an irksome, time-wasting chore. Dusting and cleaning will be made easier by new textiles which produce hardly any dust. Other domestic work could well become part of the pleasant fabric of daily life. In the same way that some people (of whom I am one) enjoy washing-up, others like housework. My own particular enjoyment is polishing brass. I could easily have replaced all my brass door-knobs with stainless steel ones if I had wanted to, but cleaning them is part of my Sunday morning routine which, in truth, I undertake for fun. If we choose, we shall be able to automate our housework, as we could our cooking, but I suspect that we will choose to spend part of our free time in the future taking a pride in doing both these domestic duties ourselves, in more or less the old-fashioned way.

The chip revolution

The impact of microprocessors, or silicon chips, will have widespread repercussions in the homes of the future in which the great range of promised domestic gadgets are installed.

Silicon chips are used to store information and instructions that have been etched on to them. Although smaller than a halfpenny piece they are capable of carrying far more elaborate circuitry than the current printed circuits they will be replacing. Their primary achievement will be to simplify, yet at the same time diversify, the operations of the machines in which they are installed. This will be true of machines as different as the cooker in the kitchen and the computerized filing system in the office.

In the home, computerized systems will be employed to operate as many household devices as we care to adapt to microprocessor technology, and the ultimate goal of this technology is to produce machines that will respond to the human voice.

At present there are prototype lighting systems that switch on, dim and switch off when activated by a number of staccato sounds, like a hand clap, beaten out in a code. These are the first in the line of audio-controlled systems that will ultimately switch on, grow bright, dim and switch off when simply asked to do so. The same technique will be applied to other operations in the home – drawing curtains, opening doors, opening windows, switching on heating systems and, of course, giving instructions to the home robot and the home computer, described later.

Outside the house the human voice might be the only means of access into the garage or the house itself. The nerve-centre of the home, the home computer, will monitor domestic security against fires or break-ins as well as the everyday operation of the heating and lighting systems. On leaving the house, doors and windows will be automatically locked. In the case of unauthorized entry the computer will sound the alarm in the police station to which it is linked and probably lock the intruder in whichever room he happens to be at the time of sounding the alarm. On the other hand, when the householder returns, he will only need to speak into a microphone near the front door to gain access. Provided that his voice tallies with the 'voice-print' stored in the computer's memory, the door will be opened without the need to use keys that can be lost or stolen by burglars. This is the same principle of voice identification described later in connection with data security.

As far as domestic machines are concerned, I have already mentioned the developments forecast for the dishwasher and the home-robot. In the case of another common household device, the washing machine, the incorporation of silicon chips will give far greater control to the user over the actual washing process. The machine will be able to advise on the quantity of powder that should be used and it will select the most suitable water temperature, according to the fabrics to be washed, simply as a result of analyzing coded labels attached to each article as it is fed into the machine.

We may even be able to buy microprocessor-controlled

furniture in 2030. Chairs and settees might be built so that they can mould themselves to the contours of our bodies as we sit or lie in them, providing the optimum comfort and support whether we are watching three-dimensional television lying down or sitting upright playing chess with the home computer.

So in the evening we may be relaxing in the living-room, or 'leisure centre' as it may have become in fifty years' time, snuggled into the electronically-contoured seating. The home computer will have been instructed to close the windows and doors, to switch off any lights that have been left on unnecessarily, and to draw the curtains and boost the heating in the bedrooms. We will be able to select our evening's entertainment by merely saying *Stagecoach* or *Match of the Day* and the old film or the football game will appear on the television screen. Though these ideas sound fanciful they are more than just idle dreams. Whether we choose to install such equipment in our homes in fifty years, or whether we do not, it will be available to us, thanks to the millions of tiny pieces of silica crystal, the revolutionary silicon chips.

Lifestyle 2030

Tired nature's sweet restorer

Of all the physical processes that support life, sleep is one of the most important. It is the period of regeneration for mind and body, the means whereby we mollify the traumas and anxieties of yesterday, to prepare and fortify ourselves for the traumas and anxieties of tomorrow. A night without sleep leads to a day without sunshine, to mix the proverbs. Of all the bodily functions we undergo during our lives, sleep is the one perhaps most intimately connected with our futures.

> . . . the innocent sleep,
> Sleep that knits up the ravell'd sleeve of care,
> The death of each day's life, sore labour's bath,
> Balm of hurt minds, great nature's second course,
> Chief nourisher in life's feast . . .

says Macbeth, a man who above all others was paranoiacally concerned with what the future held in store. Our mien, our mood, the manner in which we see the world and the world sees us, is directly related to the way we have slept. In spite of this vital role, however, there are some who seek to burden those blissful hours in the land of nod with even greater demands and responsibilities.

Many experiments have been conducted into the use of sleep as an effective medium for learning. Students have been tucked up for the night with ear-phones clamped to their heads through which information has been imparted to their tired brains in an effort to see if they are capable of retaining this subconsciously acquired knowledge. Others have been given something to learn immediately before falling asleep to test their powers of sleep-memory against their peers who have remained awake. The sleepers have come out on top. Basing their views on the apparently successful outcome of experiments like these, the proponents of sleep-memory look forward to a time when we shall all be coaxing our exhausted minds to saturate themselves with eight-hours-worth of information every night. Dreams will be a luxury afforded only to those who are on official vacations or those rated incapable of rational thought when awake; for them, ignorance will indeed be bliss. Examination revision, business reports, even important telephone numbers could all be injected into us at night if these ideas catch on and sleep loses its 'innocence' and becomes truly 'knowing'.

We have to be careful, however, not to confuse learning with cramming. When I was at school, I was a good examinee because I was good at cramming. I would wake up at about half-past six, as I still do, I would get out my Latin grammar or my arithmetic book, which I no longer do, and I would lie in bed learning until breakfast time. Then I could sit down at the exam three hours later and reproduce whole pages of information I had crammed into my head. This was not learning.

There are other factors to be taken into consideration as well, like the attention factor. There was a celebrated experiment conducted on workers on an assembly line putting together electrical components. First of all they listened to music while they worked and their efficiency rose. Then the level of illumination was raised and their efficiency rose. Then the temperature was allowed to fluctuate, so that it was cool one moment and warm the next and their efficiency rose. Then they were given a special nutritious biscuit during the tea breaks and their efficiency rose. The industrial psychologists conducting this experiment were overjoyed, the assembly workers were performing more and more efficiently. They therefore decided to trim the cost by turning the lights down again, and the efficiency rose again! All that was happening was that the workers were stimulated by the attention of all these learned people watching what they were doing. I suspect the same is true of the 'guinea-pigs' who took part in these learning experiments in their sleep.

Perhaps in A.D. 2030, electronically assisted sleep will be the vogue. Electrodes attached to the skull will pass a small electric current through the brain to stimulate sleep. Hours of restless tossing and turning, drinking malted milk and taking sleeping pills will be replaced by the simple flick of a switch and we shall literally 'go out like a light'. In the morning, the automatic time-setter will switch off the current and we shall come to, fresh and alert like a newly-charged car battery.

Nevertheless, attitudes and measures of this sort suggest a fundamental lack of humility and understanding of the human condition. In fifty years' time I believe we are going to continue to be burdened, as we have since the beginning of time, by human fallibility; and one of the ways in which this afflicts us is worry. If you go to bed knowing that you have been found out in what you have done wrong, that your bank balance is in the red and that your wife has gone off with the milkman, naturally you are going to worry and this will prevent you from falling asleep. In the year 2030 you are still going to lie awake half the night wondering whether she will come back to you or if your boss will discover that you have sneaked the tea-money and spent it all on riotous living.

As a cure for these worries, the present misguided belief in the power of sleeping pills is already showing signs of waning. I believe that the great discovery that will be made about sleep in the future is that the cure for insomnia is to go to bed later. If people go to bed later, when they are really tired, they will sleep later. So instead of going to bed at 10 p.m. and waking at 6 a.m. thinking they have insomnia, they would go to bed at 12 p.m. and wake at 8 a.m., happy that this is the 'proper time' to wake up.

In fifty years, doctors will not prescribe barbiturates for people who complain that they cannot sleep. Instead these people will be made to realize that sleep is part of human physiology, not some arbitrary process that we can turn off and on at will. I hope that our wisdom will have increased by then so that more and more people will come to appreciate that clever sleeping pills and other forms of artificially stimulated sleep do not help us. In fact, they somewhat blunt our intellectual accuity the following morning, no matter how clever we are.

Let the sleepers awake

Automatic home-control systems activated by solar panels on the roof that play sweet music at our pre-set waking time, turn up the central heating, cook breakfast and draw the curtains in the bedroom could be a reality in future homes if the dreams of the gadget-makers come true. We might even be woken from our electronically assisted sleep by a gentle prick from our wristwatches or a less gentle mechanical arm swinging a damp cloth across our unsuspecting faces.

Once awake we would summon the home robot from the landing by pressing the appropriate button on the console beside the bed, and instruct it to fetch the now-cooked breakfast. Breakfast in bed could become a daily event, not some rare treat reserved for Mother's Day or visiting elderly relatives. The wonders of micro-technology will end the chilly dash along the landing to the bathroom first thing in the morning and could remove the need to get out of bed for this purpose altogether if some way were found of electronically regulating natural physical needs. Providing, of course, that the electricity supply is not cut off. If this happens we might not even wake up in the first place. Perhaps it will be prudent, even in 2030, to keep an old-fashioned clockwork alarm clock just in case.

Although working hours will be shorter, I foresee people getting up at much the same time in 2030 as they do today. One change I would like to see happening is in the way we regulate time during the year. Instead of putting the clocks forward an hour in the spring and back an hour in the autumn, what I would prefer would be a system in which the regulator in our watches would be moved forward in mid-winter, so that we could gain during the spring months until Midsummer day. Then watch-regulators would be put back an hour until the middle of winter. This process would maximize the benefits of daylight saving. Governments in the next fifty years, anxious to economize on energy, could encourage us all to wear computerized watches programmed to move their regulators at the correct time so that we all keep synchronized. On the one day of the year when all the clocks are right there could be a great festival. This could be called Christmas.

As for early morning tea, which some gadget-makers would no doubt like to be prepared on a super twenty-first century teasmade, I should like to think that it will have disappeared in fifty years. Housemaids and butlers no longer tap on our doors to offer us this beverage, and in the future people will lose the habit of slopping tea about in a saucer when they are half awake.

Teething troubles

One of the first things most people do in the morning is to attend to their teeth. For those fortunate enough to be in possession of one of the two sets God gave them, this attention takes the form of brushing with toothpaste; for other people there is the removal and cleaning of dentures. There are ideas current in the world of orthodontics that in the future most of us will be fitted out with false teeth made from a form of a carbon or a totally new metal, which will be rooted into our jaws. Surgery of this type is already available to a minority of wealthy guinea-pigs, but in the future apparently we can all expect to be fitted with artificial teeth that will form an almost natural bond with the bone in our jaws, when our natural teeth fall out as a result of tooth decay. Similarly, some dentists look forward to the time when our natural teeth will be capped by a special plastic substance to protect their enamel and prevent it from chipping. Whatever measures are taken in the future, dental caries (tooth decay) is almost certain still to be affecting us fifty years from now. Dental caries is an infectious disease caused by micro-organisms which break through the dentine in our teeth and infect our gums, rotting the bone in which our teeth are set and causing them to fall out.

Tooth decay has afflicted man since he lived in caves. The skull of Zimbabwe Man who lived 100,000 years ago had ten badly decayed teeth. Ninety-seven per cent of American schoolchildren today suffer from some form of tooth decay and roughly four tons of decayed teeth are pulled from the mouths of schoolchildren in England and Wales every year. During the Great Plague of London in 1665 the official death figures show that after the plague itself the second highest number of deaths were caused by what was ambiguously described as 'teeth'. Tooth decay has been with us for too long for it miraculously to disappear in a mere five decades.

The way in which we will actually be cleaning our teeth, though, is an interesting topic for speculation. We tend to think that cleaning our teeth is a scientific operation. We buy tubes of toothpaste that contain special chemical formulae designed to 'fight tooth decay' and carefully rub these into our teeth and gums. Yet when we next visit the dentist he always manages to find something to fill or extract. In fact, the best way to clean our teeth is with a chewing stick which requires no toothpaste at all. Brushing one's teeth with salt would clean them as effectively as does toothpaste. In the main, toothpaste is merely an agreeable and non-fattening form of confectionery. The fact that our mouths feel fresh after brushing our teeth does not really mean that we have cleaned all the residue and traces of food out from the gaps between them. It merely means that the toothpaste contains ingredients that give us a fresh, cool feeling. In the future, toothpaste may well continue as an aesthetic lubricant to the process of brushing; its taste may differ but the principle will probably remain the same. The real cleaning is done by the agitation of the brush. Cleaning your teeth is like sweeping the carpet, to be effective you have to get at the little bits stuck at the bottom.

Throughout history men and women have been conscious of the effect on their appearance that the loss of a tooth can produce. The Etruscans were making dentures 700 years before Christ. Queen Elizabeth I used to pad her face out with cotton when she appeared in public because she had lost her teeth. There was even a thriving trade in teeth taken from corpses left on battlefields in the nineteenth century. They were used in sets of dentures. The great advance in false teeth in the next fifty years will be the development of self-cleaning dentures. These will be hydrophobic. Like the glass in the windows and the material used to make the bath and the sink, the dirt will simply wash off them.

The Sweeney Todd complex

Great efforts have been made to improve the laborious business – to which clean-shaven men must submit themselves – of shaving every morning. These efforts will continue to be made to make the procedure quicker and easier in the future. It has been suggested that some benefit could be obtained from experimenting in hormone chemistry, with the idea of preventing facial hair from growing at all. The trouble here is that the growth of facial hair is so closely linked to one's secondary sexual characteristics that where you might win on the roundabouts you would lose on the swings, so to speak. After all, it would be over-reacting if, merely to avoid the trouble of shaving, one found that one had become completely bald and lost one's sex drive.

The use of depilatories so far has not been found to be a satisfactory substitute for shaving. I doubt whether it will have become so by 2030.

Some work has been done on the chemical shearing of sheep, whereby the animals were fed a depilatory substance when their fleeces had grown to the required length. Then instead of shearing with mechanical clippers and manhandling the reluctant animals, they were simply grabbed by the tail, given a sharp tug, and all the fleece came off in one piece. But this was not found to be very effective since the sheep had a tendency to die. I suspect, therefore, that beards are going to grow and that men who want to shave them off will continue to have to do so.

Where there is going to be an improvement, in my view, is in mechanical shaving devices. History has shown that successful innovations in shaving have been immensely popular. The first year of trading for the American Safety Razor Company showed disappointing figures, with sales of fifty-one razors and 168 blades. The following year Mr Gillette and his colleagues were understandably more satisfied. The figures had risen to 90,000 razors and 12,400,000 blades. Seventy-five years later this method is still popular because, though troublesome, it is more efficient than existing electric razors; but then the 'cut-throat', so dexterously wielded by Sweeney Todd, held the safety razor at bay for many years.

I expect that before fifty years have elapsed, the clever Japanese will invent a mechanical razor based on the principle of the electric carving knife, which is so efficient that when using it to carve meat on a wooden dish, you are in danger of cutting the dish in half while aiming for the Sunday roast. An electric razor that worked as efficiently as this would sweep the board.

2030? I won't have a thing to wear

The one thing on which all the crystal-ball gazers agree is that fashion is unpredictable. In the next fifty years fashion will go backwards and forwards in history as it has always done. We shall see bright colours, shirts open to the navel and jewellery for men come and go. And the same for women. There are advantages in dressing as they do in the People's Republic of China in cotton dungarees and combat jackets. It is inexpensive, practical and convenient. It would also be convenient if everyone wore one simple outfit like Sir Winston Churchill's wartime siren suit. This could be fitted with a washable inside lining and would thus avoid the need for shirts, blouses or underclothes. However, clothing is only partially functional, some protection to keep out the cold and the wet. To an equal degree it possesses aesthetic and social dimensions.

Clothing which you put on the outside of your body is like food which goes inside. At its lowest level, food provides the chemical inputs that keeps the body's machine going, just as clothing at its lowest level is put on to cover you and keep the weather out. But food at its next level is aesthetic, you eat it because you like it. Clothes are also worn to elicit comments like, 'What a beautiful dress, darling!' The third dimension for both food and clothing is a moral one. On the northern side of the English Channel it is not considered fit to eat horseflesh but on the south side it is widely eaten. Where clothes are concerned, most people who are going to be presented at Buckingham Palace would consider it fitting to wear morning dress. There are convinced socialists who refuse to wear this costume because they feel it immoral to do so. However, neither group would think it right to go in overalls. Fashion itself is influenced by so many extraordinary social influences and factors that it is difficult to foresee what course it will take in the twenty-first century.

I can conceive, though, that fifty years might be just the right time to see more formality in dress. It could be more generally realized that by adopting a code of dress, social life becomes easier and people find that they 'know where they are'. It has been said that one generation's leisure wear is the next generation's formal wear. Dinner jackets were originally only worn in the absence of ladies. The old stiff-fronted evening shirts have been replaced by nylon soft-fronted ones which drip dry over the bath, instead of having to be laundered and starched every time they are worn. The fashion of wearing polo-necked evening shirts came and went. So fashions will change. Who knows, black jackets and jeans may become the formal dress of the 2030s.

I anticipate that our clothes in the future will represent our attitudes to the different phases of our lives. With the great increase in leisure people will wear leisure clothes as never before. But these will become more specialist as we become more affluent. For example, many more of us will want to buy and wear special outfits for gardening. These might be traditional wellington boots, cords and a pullover, or they might be smart, one-piece jump-suits with knee-patches, pockets and areas designed for rubbing hands clean. Whatever we chose to wear, our greater affluence will determine that these are our gardening clothes, as opposed to our old casual wear which has now been relegated to use in the garden.

With a return to formality in business dress I am confident that men will take a greater interest in their appearance. Tailors will become re-established as the interest in handicrafts flourishes with the result that more and more of us will choose to spend a little extra on bespoke suits in preference to the mass-produced two-pieces that we wear at the moment. The ability to take part in clothes designing, described later, will be of interest to both sexes, and coupled with more free time, the buying and wearing of clothes will once again become a pleasure and a source of enjoyment for all.

Of course there are many who look forward to the future as a time when traditional concepts of clothing design and function disappear completely to be replaced by the concept of clothes acting as a second skin. These would not attempt to disguise or decorate the wearer: their utilitarian functions would merely cover and insulate him or her. Appearance would then become a matter of physical fitness and bearing. No man would be singled out from his peers by the nature of his clothes alone, and so much of the snobbery and elitism attached to fashion today would be done away with entirely.

Wool will survive because, whatever happens to our economy and that of the world in general, we will always rear sheep as the only way of profiting from rough and hilly land. And having reared them, we might as well shear them. But apart from wool and cotton, which is easy to grow, man-made fibres will dominate the textile industry of the future. They are strong, easy to clean and infinitely versatile. It is possible to make sheer silky material that would have caused the Queen of Sheba to swoon; nothing like it was ever seen in antiquity. We can manufacture materials that look like fur for elegance or for clothes to be worn in the Arctic. Such materials are better than those made from biological materials. Fibres are being improved all the time and in fifty years they will be certainly superior to the ones we are using at present.

Improvements in textiles will produce their own changes in clothing. Hosiery manufacturers may at last produce a totally ladder-proof material for stockings and tights. Permanently creased fabrics will save the labour of ironing. New techniques of waterproofing will enable all outer garments to repel rain, without affecting the exterior appearance of the fabric concerned. Perhaps the most important consideration in the clothes of the future will be comfort. Whether we are all wearing Mao suits, Edwardian high fashion or sequin-covered body-stockings is impossible to predict, but I will be bold enough to say that whatever we do wear will be conditioned by comfort as the primary consideration.

One area in which we are likely to see a decline rather than an advance is in disposable clothes. In the twenty years or so they have been available they have not made much headway. One reason is because man-made fibres are easy to wash and dry, and are strong and durable. Furthermore, disposable clothes are not cheap.

It is a safe prediction that plastics will make progress for the manufacture of shoes. The plastics industry will finally succeed in producing a really effective substitute for leather, namely something that is strong yet able to let the feet 'breathe'. To achieve this, it will need to possess the ability to let air in and moisture out. At present, plastics used in shoes tend to crack. The cobbler to whom you take them to be repaired says, 'Sorry squire. I'm afraid I can't mend plastic shoes.' But in the future this problem will be overcome. The makers of shoes, like those of motor cars, will see the wisdom of providing spare parts and after-sales service. A further giant step forward will be plastic shoes that are permanently elastic. By this means, the nuisance of shoe-laces will be done away with. We shall slip our feet into totally elastic shoes which will fit like gloves and never leak.

Time pieces

The most intimate connection that most of us have with sophisticated technology is the instrument we carry with us to tell the time. Our lives are controlled by time. We have to get up at a certain time in order to be able to leave the house at a certain time, so that we will catch the train to arrive at work on time. A constant awareness of time is necessary if we are to remain in the mainstream of life. The watches we strap on to our wrists in the morning are our link with the 'clockworks' of life. Lose your watch for only a day and you feel disoriented and confused, even though there are clocks everywhere giving the same horological information. It is comforting to have what amounts to a microcosm of this invisible universal force attached to our bodies as a means of private reference. It helps us orientate ourselves in the world at large and remain in control of our own 'time' at least.

Given this importance that we attach to watches, many ideas have developed to improve on their performance and function. Miniaturization has enabled watchmakers to include more and more sophisticated equipment in their products. It is not uncommon now to find a watch that will tell you the time in another city on the other side of the globe, or one capable of acting as a stop-watch, accurate to one-tenth of a second. So in the future we shall be using watches to calculate mathematical problems, to communicate with computer centres by radio transmissions, perhaps even to see our friends on miniature video screens. These are some of the ideas that technological advance has encouraged us to believe that we both need and want. It is already possible to power wrist-watches with nuclear energy, using minute particles of radio-active material that give off radiation at a very regular rate. These watches can provide accurate time-keeping from birth to death, with never the need for adjustment or battery renewal.

No doubt it would be useful to have a watch that told you that you had forgotten your wife's birthday or one that could work out cube roots, but how many people actually want to know these things? Rather, I believe that the watch of the future will tell the time and that it will devote its whole operation to telling the time. It will probably be a digital watch, and my guess is that it will either be powered by a small battery or else, like today's self-winding watches, will use its self-winding mechanism to charge a small battery. I have a feeling that modern digital watches are becoming so crowded with information that it is difficult to read the time. Perhaps in fifty years from now our watches will tell us the date as well as the time and will probably incorporate an alarm, but cube roots should by then be omitted.

Where there will be a major advance is in the aesthetics of the watch. The current digital watches are ugly. They are big and clumsy and lack the delicate craftsmanship that used to be associated with watchmaking. In the future, far greater attention will be paid to the appearance of watches, as I hope it will be paid to the appearance of ceramics and kitchen furniture. I have no objection to a watch telling the time accurately, that is its principal function, but it is also a comment on the personality of the wearer, and it would be very pleasant to see a return to watches and time-pieces of every sort that were both functional and elegant. They would not only be a statement on their wearers, but they would also show an aesthetic awareness in society as a whole, an awareness sadly lacking in our present computer-dominated age.

Mens sana in corpore sano

'Your prayer must be that you have a sound mind in a sound body,' wrote the Roman poet Juvenal nearly 2000 years ago and his advice will be as important and applicable in the twenty-first century A.D. as it was in the first. By careful use of our considerable leisure time we will be in a position to blend the currently separated concepts of 'work' and 'leisure'. With an increase in their non-working time people will come to realize that life is a seamless garment which we wear right round the twenty-four hours of the day. Whether we are at work, at play, or resting, we are equally ourselves and what we do can be equally valuable. I foresee that in fifty years we shall have come to integrate our leisure and recreation, our thinking and our social activities together with the industrial 'employment' for which we are paid.

At present far too many people attempt to cram what they consider to be suitable exercise into brief periods of leisure between their otherwise sedentary periods of work. These sudden bursts of frantic, physical exertion can often lead to serious health risks for those who are not fit. As far as I can tell, the fashion for jogging is already showing signs of waning. It has done little good to people dying in mid-stride from heart attacks or to others who trip and fall. With more free time and shorter distances to travel to work many more of us will be walking there than today. We will be using the stairs to move from one storey to another, saving electrical energy wasted on lifts and saving ourselves from obesity and shortness of breath. The nodding interest we have at present in fitness will continue into the twenty-first century, but as now it will probably not prevent us from smoking and drinking, even though there will be unimpeachable medical evidence to prove that these habits do us harm. If the priggish minority try to impose their will to make the majority abstemious by Act of Parliament, what I call the 'Volsted effect' will come into play. The Volsted Act was prohibition in the USA which was introduced after the First World War, in the determination that homecoming soldiers should become virtuous teetotallers. It had socially disastrous effects on American society. So far, the lesson has been learned, and decrees to ban unhealthy practices like smoking and drinking have not been promulgated. This is not to say that such will not be done in the next fifty years, inviting equally disastrous effects in the 2030s as in the 1930s.

We will be feeling fitter and healthier in the future that lies ahead if we listen to reason and take fewer pills and tablets. Sleeping pills do not in fact bring extra happiness or pleasure, aspirin can easily be taken in excess, and, in fifty years, constipation pills could usefully have disappeared, too. A hundred years ago, these pills were worth a guinea a box to people who wanted to be 'regular'. Next came the 'inner cleanliness' fanatics and, after that, the drinkers of liquid paraffin. We have given up dosing ourselves with epsom salts; we have given up swallowing liquid paraffin; pills for constipation are a distant memory. I prognosticate that by the year 2030, we shall have given up taking pills for this trouble altogether for the simple reason that we will follow a nutritious balanced diet which will enable Nature to make us regular all by herself. Along with the cult of patent medicines I hope – indeed I expect – to see the similar cult of dietary supplements such as vitamins, proteins and minerals to disappear. Before 2030, it is to be expected that people's unnecessary preoccupation with their health, regardless of the fact that they are quite well, will have come to an end. Two words summarize the rules of good nutrition and a healthy diet: moderation and mixture. By following these simple precepts in the future we will be able to eat everything which we eat today and enjoy excellent health without the aid of further chemical assistance.

Medicine men

Today a visit to the hospital is a sign of medical superiority in the eyes of fellow sufferers, as important as the awesome word 'specialist' was fifty years ago. Our present-day hospitals have become microcosms of the larger world they aim to serve. They have their own laundries, their own cooking facilities, places to work, places to play and places to die. They are staffed by their own bureaucracy, they have their own hierarchies, and, until recently, their internal structure was ordered and tightly controlled. However, like the great world outside the gates, the hospital has become disordered and top-heavy. The frightening spectacle of hospital strikes that the British public witnessed in the winter of 1978–79 has become a reality that both the world of medicine and the world of politics will have to accept and come to terms with. Since it is very doubtful that ambulance men and hospital porters will ever allow themselves to be controlled by senior medical staff during industrial disputes in the future, we will have to face a situation in which hospitals could be constantly under the threat of closure or blockade. In addition, the hospitals themselves are becoming overcrowded and expensive to maintain, while the waiting lists for treatment are growing longer all the time.

By the year 2030, and probably a good deal sooner, the general public and the medical profession will demand a remedy to this unsatisfactory state of affairs. They may perhaps find the remedy in general practitioners. Doctors who are primarily concerned with the practice of medicine will be drawn, no doubt with a little political encouragement, to favour the idea of a growth in the number of general practitioners, dealing with patients at grass-roots level and saving them the trouble of going to the hospital.

A further reason for the decline in hospital treatment is that the technology of machines like the brain scanner and the mass-spectrometer is beginning to kill itself, like the mastodon, through overweight. More and more technical personnel are required to operate the machines and should these people choose not to do so for any reason, the system begins to crumble. The direction of medical technology in the future may therefore swing away from multi-million-pound machines and concentrate on equipping the general practitioner to diagnose and treat all but the most serious conditions himself.

Say aaaagh

Ideas along the lines of diagnostic machines for the doctor's own surgery are already being formulated. Some anticipate the development of health chairs in which the patient will sit while sensory devices in the chair weigh him, measure the temperature of every part of his skin, check his heart-beat, take X-rays in three dimensions, test his reflexes and see how much hair he has lost since the last visit. Other possible innovations might be monitoring devices worn by patients liable to suffer from heart conditions. These would be attached to the patient the whole time and would send constant signals to a central receiving computer. Any change in the condition of a particular patient would be immediately passed to his or her G.P. who would know their exact significance and would be able to organize emergency aid. These monitoring devices would also give long-term warning of possible heart attacks which would enable the doctor to contact the patient for a full-scale examination and take appropriate clinical action before an attack took place. It has even been suggested that every home will contain a miniature television camera that patients would use to transmit a picture of an observable ailment to the video-screen in the doctor's surgery. The logical conclusion to this idea might be to keep a miniature television camera attached to a fibre-optic cable in the bathroom cupboard. When plugged into the home television transmitter the patient could feed the thin flexible thread into either end of his or her body to let the doctor have a look inside to see what was going wrong.

Like so many fanciful schemes, it is unlikely that this one will be in general use in fifty years. In my opinion it is far more likely that the future age of microtechnology will produce a much simpler but equally useful portable machine that would act as a general diagnoser and which could be carried round in a doctor's black bag. This could be a machine that will record temperatures and blood pressures, analyze urine and take electrocardiographs, all of which have to be recorded independently at present. Diagnosis with this machine will be similar to the equipment used to tune car engines. All the clinical information will be fed into the automatic diagnoser and the results displayed to the doctor immediately. The function of this machine will be similar to that of the health chair, but the equipment, being portable, will be more useful to the busy doctor moving about his practice.

Similar developments in miniaturization and the silicon chip will bring X-ray equipment back into the G.P.'s surgery, too.

Computers will have an increasing impact on medical practice. There is one school of thought that holds the belief that patients actually prefer to talk to an inanimate machine about their physical condition and consequently reveal more information than they would do if they were talking to a doctor face to face. If this opinion is found to be correct, we may when we feel ill fifty years from now call up not the doctor, but his computer. The machine will then ask the questions: 'where does it hurt?' 'put out your tongue at the video-screen'; 'say aaagh'; 'have you got diarrhoea?' 'is the pain constant or only intermittent?' The computer will store the information for the doctor to retrieve and examine in the time allocated to 'out-patients'.

An alternate possibility is for the doctor to use his computer to store his medical records and match particular cases to the recorded clinical advice of acknowledged experts in any branch of medicine. The diagnostic machine and the computer will not be depended upon to uncover all the information about a particular case. They will, however, quickly supplement what the doctor finds out for himself when he comes to examine his patient face to face.

Fifty years from now, the current myth will be exploded that, no matter what disease we suffer from, science has the cure. We shall then know that man is not God, that misfortune may strike at anyone and unknown things may happen which are beyond the cures of any doctor. If we realize this, then we will appreciate that a doctor is only a trained man or woman doing the best he or she can under difficult circumstances.

I think that there will always be a place for an alternative form of medicine, as there will be of education, because of the anarchic way that people like to break out of any organized system. One way in which this alternative practice might be achieved could be, as for plumbing or painting the house, in the growth of do-it-yourself medicine. I can easily foresee this complementing the elaborate national health service and the expensive private sector. Quite soon we may begin to see the publication of knowledgeable books on 'diagnosing for yourself' and 'curing for yourself'. These will tell you that if you have flu there is no need to go and bother the doctor when all you have to do is to go to bed. Likewise if you cut your finger and it becomes poisoned the book will tell you that you need antibiotics which you can get from a chemist. At present the chemist will not sell you antibiotics without a doctor's prescription, because he is frightened that you will by mistaken treatment make your infection resistant to the antibiotic. But informed people are not such fools as doctors think. Such an attitude denigrates the intelligence of the general public. In the next fifty years, I see an awakening to the fact that ordinary citizens are intelligent people capable of following instructions. Faith in responsible citizens could radically lighten the work load of the health service and do-it-yourself medicine is another GASH that will equip the individual to maintain his independence if he so chooses. He will have to be prepared to take the risk that there may be something very wrong that may kill him, rather than wait to see a doctor who could be wrong too. He makes the same choice when he decides not to wear his seat-belt when he is driving, although statistics will tell him that he is more likely to get himself killed that way. We will grow accustomed to do-it-yourself medicine which could be another leisure activity. 'How interesting it is,' we will tell our friends, 'I'm curing my acne.'

The Shangri-la complex

Elixirs of life, secrets of eternal youth and ascetic regimes of living have all failed to prolong significantly the span of human life beyond the biblical 'three score and ten' years. The methods employed have often been as bizarre as the results claimed for them. Virgins' blood has always been a favourite. Chinese emperors swallowed pills made from this elixir, while their eunuchs also secretly consumed them in the vain hope that they might restore their lost capabilities. Both groups were sadly disappointed. Then there was the Polish countess, Elizabeth Batharoi, who put a rather too literal interpretation on the phrase 'royal blood'. She was found guilty in 1610 of murdering 650 young girls in order to take regular baths in their warm blood, in the hope that this would renew her youth. The only effect it had on her declining years was to have her incarcerated for life in her castle. More modest expectations fill present-day minds that contemplate the prospect of their own demise.

There are those who have opted to become guinea-pigs of cryobiology, the science of the effect of freezing and low temperatures on living organisms. Immediately after death their bodies have been deep-frozen in baths of liquid nitrogen where they will remain until such time when medical science has found a cure for the condition that killed them. When this happens they will be thawed out and, they hope, restored to life in order to be cured. This could be in several centuries time. One may ask how will they be able to pay for the treatment when they are thawed out? It would be a terrible anticlimax to wait all that time to find that in the end they could not afford to be resuscitated.

Others with less grandiose expectations, who want to enjoy for as long as possible life with their family and friends, can arrange for defective organs to be replaced by 'spare-part' surgery. There are many people walking about today thanks to the development of artificial hip joints. Titanium skull plates have produced remarkable results for people with severe head injuries. Dentures have been widely used for millennia. With the ever-accelerating advance of technology, the hopeful are looking into the next century to see what other parts of the human anatomy can be effectively replaced by artificial devices or machines. We may even be able to replace organs with the same degree of success in the future with which we are replacing joints at present.

The organ most liable to failure is the heart. Heart disease of one form or another is a major cause of death in the developed world where increasing numbers survive to old age, and the invention of an efficient artificial heart could prolong the lives of thousands of people every year. Experiments on animals have led to the production of successful blood pumps which in turn have given rise to the design of a heart (effectively two pumps) made from smooth artificial rubber. The great problem that has to be overcome before any artificial heart can be confidently installed is how will it be made to work year in year out for perhaps fifty years?

There is one plan to use nuclear energy to power artificial hearts in the future as is already done with 'pacemakers'. These would contain a fuel capsule of radio-active material, a means of converting the energy to drive the pump (heart), and a computerized mechanism to regulate the rate of action of the heart in accordance with the body's needs.

The artificial kidney might be miniaturized in fifty years time to the size that would enable it to be carried about in a briefcase or handbag. There is already a machine in existence that reads the printed word aloud to the blind and many anticipate that there will be devices in the future enabling them to 'see', or to recognize images by means of sense vibrators activated by a miniature television camera. So even if we do not succeed in finding the answer to eternal youth, fitting ourselves out with a kit of 'spare parts' could prolong active life.

A more promising and easier way to longer life might be started right away with today's infants, who will be fifty or so in the year 2030. Experiments have shown that rats can be made to live longer by limiting the amount of food they are allowed to eat during their infancy and 'childhood'. If human babies grew the same way as rats do – which they do not – all we should need to do to lengthen their lives would be to underfeed them. If only science were so simple!

Life in the future will be extended for some of us by the elimination of certain hereditary diseases. The technique of amniocentesis has been perfected to examine the cells of a foetus in its early stages of growth. A sample of the amniotic fluid surrounding the foetus is removed from the womb. Cells contained in the sample are then grown in a tissue culture to be examined for evidence of abnormal chromosomes or the presence of specific enzymes. In the event of the discovery of a serious genetic disorder, the foetus is aborted. The use of this procedure could well mean that certain hereditary diseases will be eliminated in the future.

Whether we live three score and ten or five score and ten years, we all have to die. In our present clinical age we deny death. Most of us depart this life, to quote T. S. Eliot, 'not with a bang but a whimper'. We lie in unfamiliar surroundings, dressed in unfamiliar clothes, with pipes stuck into all our orifices – case numbers waiting for a specialist to decide whether the time has come to switch off the apparatus that has kept us clinging to life.

My own view, which I know is quite contrary to the thinking of many others, is that in the future, fifty years on, recognizing the inadequacy of our present situation, we may well decide to die at home. But if indeed we so decide, there are things to be done. The best bedroom must be redecorated especially for the occasion (if there is a best bedroom), invitations must be sent out to friends and relatives (if there is anyone to send them), and we must be dressed in our best pyjamas (if there is anyone to dress us). General Eisenhower regularly used to wear pyjamas with the five stars of his rank on the lapels, so there is no reason why the ordinary man should not be allowed something rather dashing in silk to see him off in his moment of triumph. Death could be a moment to be savoured. Death can be sad, it can be sordid or it may be horrible and humiliating. This, however, is no reason why, fifty years from now, we should not try to make it better than it is today.

The Dignity of Labour

Work

Apart from the technological changes which will radically alter the pattern of work in the future, the nature of that work and our attitudes towards it will undergo equal changes. Moving from an industrial society to a post-industrial society will not only require coming to terms with wide-scale automation and increased leisure-time, it will also mean adopting new priorities in human work.

The Industrial Revolution was characterized by the growth in the manufacturing sector of the economy at the expense of the agricultural sector. With the evolution of the post-industrial society the number of people employed in manufacturing industry will decline and the service industries will grow in importance to become the principal employers of manpower.

No doubt a fully automated industrial economy is more efficient than one that relies on employing a human labour force. Those industrial nations which have invested or which are investing in new automated plant and machinery will not only be the industrial leaders of tomorrow, they are the industrial leaders of today.

In the British economy there are many experts who are convinced that it would be better to concentrate right away on the high technology of the service sector, in which Britain has always been in the forefront, rather than attempt to compete with the advanced industrialized nations who are already a decade ahead in industrial automation. In their opinion the future of the British economy must lie in providing highly trained skills in computing and information technology, not in the production of competitive manufactured goods like motor cars and television sets.

Whatever the eventual direction of individual national economies, work itself will be very different in 2030. Those who peer into the future through the rosiest of tinted glasses see a state of national prosperity in which every citizen will receive a basic national minimum income from the government, tax free. This income would be adequate to maintain a perfectly agreeable standard of living, but it would not be sufficient to permit the recipient to indulge in the luxuries of future life like trips to the moon or collecting veteran cars from the 1980s.

The individual would be free to choose whether he or she wished to spend all his or her time living on this income alone and benefiting from the unlimited free time it granted them. The alternative at the other end of the economic ladder would be to perform some disagreeable or unpopular work for which society would be prepared to pay very high wages. In between these two extremes would fall all other forms of employment, which would be reimbursed according to their popularity and interest. Jobs that might be considered popular and interesting would only command relatively low wages compared with those which few people would be prepared to undertake because of their disagreeable nature. In the future jobs of this category would be very highly paid.

These same pundits also foresee far greater choice in individual working hours than most of us are permitted at the moment. The rigid timing of the factory whistle and clocking-on would be replaced by a system of universal flexitime on the lines of systems operated by one or two firms at present, but applicable to the whole labour force in both factories and offices. Employees would be required to work for a minimum number of hours during a set period, perhaps a week or a month. However, they would be free to choose the actual hours in which they worked, allowing them greater control over the way in which they organized their lives. The operation of a 'credit' and 'debit' system would enable each employee to build up periods of free time that he or she might wish to take in a block, by working more hours in a set period than the agreed minimum. Alternatively it would be possible to 'owe' time which would have to be made up during the following period.

Those who wanted to finish work early would start to work early, while others who wanted a long lunch break might prefer to work extra time at the beginning or end of the day to facilitate spending two hours in the pub in the middle.

It seems more likely in my opinion, though, that an increase in our leisure time will profoundly alter the balance of our lives. Leisure will become the norm and work the exception. The concepts of job satisfaction and flexitime are based on the assumption that people are essentially dissatisfied with their employment. It presupposes that workers will seek to improve their lot either by gaining greater material reward or by exercising greater control over the way they apply themselves to their employment.

However, since the situation in the future will be fundamentally reversed, I foresee work becoming more and more popular. People will put themselves out to get it. They will not be concerned with job satisfaction in the way this meaningless phrase is bandied about today. The satisfaction will come from having a job at all. In the same way, the delight in having a job will more than compensate for the possible inconvenience of working from ten o'clock in the morning until four in the afternoon. In any case, a system of totally universal flexitime would produce so many technical and administrative complications that it would probably be impossible to operate.

Take a letter

The impact of the microtechnology of the silicon chip, already described in the home, will find its major application in the fields of communications and information storage and retrieval. Collectively these are referred to as *word processing*.

The great advances that we will be witnessing in the development and sophistication of microprocessors during the next fifty years will herald the age of universal computerization. Even in their comparatively short lifetime computers have changed in size, function and cost at a faster rate than any other machine. These changes will be accelerated as we move into the next century. Silicon chips will become capable of holding one million words or more, a fact that will reduce the size of the computers in which they are fitted while greatly increasing the amount of information that can be housed in them.

In the field of employment those jobs most directly concerned with word processing at present will be the ones to feel the greatest effect of the introduction of this new technology. Indeed, some large corporations already use certain kinds of word processing machines. The two jobs most likely to be fundamentally changed by word processors in the future are typing and filing.

The actual size and appearance of the word processors of the future will be determined by the amount of human contact with the machine. For though the computer could be reduced to a minute size, if any human contact is required (such as the operation of a keyboard or the reading of data with the naked eye) the machine must be large enough to accommodate the use of clumsy human fingers, or be capable of producing readable print.

The first generation office word processors might be slightly larger than existing typewriters, since they will be operated by a conventional keyboard. They will probably look like a cross between a typewriter and a television set. Whether the machine is being used to type a letter, or a manuscript, or whether it is the terminal for a computerized filing system, its operation will be essentially the same.

The operator will sit at the machine and will type into it the text of the letter or document, or the nature of the information to be retrieved from the filing system. Instead of the text appearing on a sheet of paper, as it does on existing typewriters, it will be displayed on the screen above the keyboard. This will provide the operator with the chance to delete or add material as an afterthought. It will also permit the recall of previously typed material stored in the computer's memory. By this means the operator will have complete freedom to edit material during composition, without needing to interfere with the finished text. Such machines already exist in printing houses which use computer typesetting.

For unlike existing typewriters, future word processors will act as editorial intermediaries between the operator and his material. The operator will be able to modify his material as he progresses, but so too will the machine. It will automatically correct errors in spelling, punctuation and grammar. It will also space the material correctly. For example, in text destined to be printed, it will be able to set the copy in the exact layout necessary for the printed page. Only when the operator has completed his text will the machine undertake the second of its functions.

Having finished typing a letter, for example, the operator will key in the address of its destination and will instruct the word processor to dispatch it via landline or satellite to the word processor at the other end. The machine will then transfer the letter to the memory of the receiving machine in a matter of seconds. There it can either be immediately displayed on the screen, or stored to be read later.

If the operator's material is destined for permanent storage, as in the case of updating records, he will instruct the machine to put it on permanent record in the computer memory.

If, on the other hand, the word processor is used to gain access to the data in a computerized filing system, the operator will instruct it to carry out a thorough search of all the material and to extract from the records all that is required in answer to the enquiry. This material will either be displayed on the screen or it will be available on a paper print-out.

Of course, in machines like the one described above, an operator is still required to perform the actual typing. This might well be the originator of the letter, or the person who will be making use of the information from the files. However, it might equally be an employee specifically engaged to operate the machine, like the present-day copy typist. But, when the ultimate goal of manufacturing voice-controlled computers is achieved, this operator will become superfluous.

Once the only form of human contact with the machine is through a microphone and a screen, the word processor will become smaller, freed from the burden of having to carry a keyboard. We will find the machines conveniently situated on desks, occupying about the same amount of space as a present-day telephone. This reduction in size and space will result in the individual offices of the future conducting the same amount of daily business and storing the same amount of information that today has to be housed in a sizeable tower-block. The whole of the administrative section of the United Nations might one day be condensed to a size that would occupy no more than one floor of its present headquarters in New York, and to run the entire organization, it might employ the same number of staff currently working on just one floor. However, these predictions must be tempered by the knowledge that many people are viewing this kind of technological change with great caution because of the massive unemployment it would cause.

An exciting extension of the use of storing information on microfiches is the likelihood that the microfiches might be readable by both men and computers. As well as recording words, diagrams, figures and pictures which will be read by the human readers, these future microfiches will also record the same information to be read by computers by means of holography, three-dimensional laser photography. The computer's information will be far more compact than a man's. Two and a half million items of digital information will be recorded on to six and a half square centimetres of microfiche. As a result of this technique, computers able to 'talk' to each other will also rapidly exchange information when the appropriate microfiches are fed into them. This means that a human reader will be able to retrieve a microfiche from the master filing cabinet, read it on a conventional scanner and then transfer the data recorded in the hologram section immediately into the computer, without the need to transpose or process the information himself.

It is possible that even microfiches may be superseded by another rapidly developing means of storing information, the so-called bubble memories. These are one or more strings of microscopic pockets of magnetism, stored on a thin film, with an individual bubble diameter of 0.0127 millimetres. Data are stored and retrieved by injecting a bubble at one end of the string, which moves the whole string along one space. As they pass a fixed point, the bubbles are 'read'; the presence of a bubble registers a digital '1', the absence a digital '0'. Due to their tiny size, it will be possible to store vast amounts of data in the bubble memories of the future, and projections speak of two million words being compressed into an area only slightly larger than a postage stamp.

Unfortunately one of the greatest disadvantages of using computers in business and finance is the security risk they represent. Spectacular 'robberies', both fictitious and actual, have been perpetrated through the perfectly correct but quite illicit re-programming of computers controlling business finances. Naturally, a prospective thief requires expert understanding of computer systems and their operation, but with the increasing use of computers this expertise is spreading. If we reach the stage when interactive, fibre-optic communications enable us to have direct access to central computers through any number of outlying terminals, it will be even more important for the computer to establish the legitimate identity of each caller, before divulging classified information.

In the future, computers will need to demand evidence of individual identity by any number of means. Identity by voice exists already in those systems that obey verbal instructions, which have been mentioned earlier. In this case, each operator dials his code number into the computer and then speaks into it to match his own voice with a pre-recorded sample. When the computer is satisfied that the two voices are the same, the operator is free to proceed with his work. Another means of identification might be the use of signatures. Perhaps the signature of the future will be even more carefully scrutinized than the ones we hastily scrawl across cheques and parking fines today. We might be required to write our signatures on a special sensitive surface that will measure the time we take to write individual letters and components, apart from comparing the form of the signature for authenticity. Even though the most skilful forger might develop the technique of producing an identical signature, it would be impossible for him to reproduce the same writing action, nib pressure and time taken to write it. A simple check with a pre-recorded signature would leave the computer in no doubt about the identity of the caller.

A more original but equally individual means of identification would be a 'breath-print', or more correctly, a gas-liquid chromatograph of one's breath. Our breath contains carbon dioxide, water, and a whole variety of other chemical compounds. In each one of us the quantities are different, making our breath as personal as our fingerprints. We might breathe into a type of microphone so that the computer could analyze our breath and say, 'There's not enough propyl alcohol for you to be Mr Smith – go away.' Perhaps as a final resort we will just stick our thumb on to the video-screen and let the computer do a fingerprint check. To avoid messiness, it would be necessary to clean the screen each time.

The current craze which has gripped us about the possible uses to which computers will be and can be put has led many of us to overlook the fundamental purpose of using computers in the first place. Their primary objective is to act as highly efficient information storage and retrieval systems. Far too many people, excited or terrified by images such as Hal the computer from *2001*, believe that in fifty years time computers will be used to determine company and even national policies. Ideas like this give rise to illogical fears that somehow computers will 'take over' and control our lives. While this is no doubt excellent material for science fiction, it can bear no similitude to actual events now or in the future. The role of computers in the society of the twenty-first century will be exactly the same as the one they play in our present semi-automated one. They will make available as much information as they possess to those who will be responsible for making decisions in the boardroom and in the cabinet. But actually making decisions will be as difficult then as it is now.

It will depend on the good judgement of the board of directors or ministers, and that of the managing director, or the Prime Minister.

Even with all the information that will be at their fingertips in 2030, senior managers will be just as likely to guess right or wrong as their predecessors today, because guessing is an intuitive process, not merely a rational assessment of facts. If they have more facts at their disposal, their information will be more accurate and they may be stopped from doing some of the idiotic things that they would otherwise have done had the information not been available. On the other hand, this increase in 'knowledge' may equally prevent them from doing the enterprising things which have transformed the fortunes of successful companies in the past. How did EMI know that the brain scanner was going to be a success or a failure? How did Rank, who were in the film business and the flour business, know that the copying machine was going to save them? Someone had to guess right, and managers will have to keep on guessing right if they are to survive in the future.

Similar enthusiasm has greeted the possibility of the use of interactive, video communications systems in sales meetings and boardroom discussions. Instead of busy directors being required to travel long distances to attend a board meeting in London, for example, the inter-firm satellite communications network will allow them to participate in group discussions and meetings while sitting in the comfort of their own offices around the world. The justification for these global meetings is that they make it easier for frequent consultation and discussion, and save expensive travel all over the world in pursuit of sales.

What the proponents of these ideas fail to realize, however, is that communication takes place in a subtle, complicated way. It is not simply a matter of speaking words. There is also communication with the eyes and what is loosely described as 'body language'. I have been taken to task for waving my arms about on television, but what some viewers fail to understand is that this is subconscious non-verbal communication. The reason why accused persons are tried before a jury is firstly to allow those twelve men and women to interpret the spoken evidence and secondly to analyze the accused person's demeanour and conduct to see if he really is guilty or not; this last process they do as best they can unconsciously.

If you want to sell a university to the Saudi Arabians, or a factory to the Chinese, you have to go and see them face to face. If you want to convince your fellow directors that their current policy of making ball-bearings is wrong and that the company should be making glass eyes instead, you have to sit round a table with them and argue it out face to face.

No matter how clever the means of communication are in the future, it will still be necessary to meet people if we really want to communicate with them. I know this to be true from my own experiences on television. People watch me and hear me talking but even so they do not get the full message. The television can so easily be insincere and ambiguous. Often people who meet me in real life say, 'You're not Magnus Pyke. You're too tall.' This is just one personal example of a general situation. No matter how advanced technological communications become, real meetings between real people will still be very important, not only in business, but in every walk of life.

For the same reason, in my view, tape-recorders and video-casette recorders will never replace the one secure province of the good intelligent, human secretary, recording the minutes of meetings. Machines may provide a verbatim account of the discussions, but the basic message may be buried in the medium. The reason for this is what is known in science as the noise and the message. There is so much talk in meetings that the real meaning, the kernel of the discussion and the decisions arising from it are confused and obscured by the barrage of noise. Whereas a good secretary can cut through the ocean of verbiage and the superfluity of information to extract the message and record it in a succinct, coherent minute that synthesizes the sense of the discussion in a brief matter-of-fact statement, the computer must take it all in.

From brawn to brain – industrial machines

The gradual development of industrial automation is more than a simple substitution of machine processing for human labour. The machines themselves have become technically more sophisticated and have altered their own functions to match this development. At the dawn of the Industrial Revolution the great advance made in manufacturing industry was primarily an advance in strength. The engines and machines developed in the early nineteenth century were not intended to replace human power, they were used to increase its effectiveness, by augmenting the strength of human muscle with mechanical muscle. Even though output and efficiency rose enormously as a result, human skills and human strength were still required to operate the machines in even the simplest manufacturing processes.

As industrial automation and technology advanced the machines became in a sense more 'intelligent'. They gradually developed from the status of being auxilliary human muscular systems to surrogates of the human nervous system. Following the invention of cybernetic control and guidance systems, machines can now be instructed by a computer to produce precision components, or cut units from a piece of valuable material to ensure the minimum wastage, doing both the thinking and the processing themselves.

Engineering and other technical designs are being drawn increasingly with the aid of computers that display them on television screens. Computer graphics, as the system is called, have already replaced traditional blueprints. The draughtsman of the future will have the computer at his disposal to test the feasibility of an idea he has sketched out and to draw it accurately. He will place a sketch on to a screen and instruct the computer to analyse the design systematically, testing it against all the known data relating to the subject in question. Where the draughtsman's ideas fail to obey any physical laws and properties, the machine will adjust them until they do. In the end it will either reject the idea as being impractical or it will produce a drawing of its own on the video-screen incorporating all the necessary modifications that will be needed to transfer the idea into a practical reality. In this case, the computer will be offering guidance and instruction at a highly technical level to the human operative. It will be extending his thinking strength, just as the earlier machines extended his muscular strength. Before industrial processes can be fully automated, however, machines will have to be equipped with additional 'human' skills and sensitivities, such as the ability to correct their own errors, and substitutes for a worker's judgement, skill and memory.

Some industrial processes are better suited to total automation than others. In certain mass-production processes where uniform commodities are assembled on a production line we shall see total automation during the next century, with a central computer overseeing and controlling the entire manufacturing cycle from start to finish. The production of motor cars is already becoming highly automated in Japan, where robot welders assemble car bodies and join seams together without the need for human participation. Current production-line procedures have reduced the assembly of cars to a series of simplified operations that can be undertaken by the machines. Many of the production machines of the future exist in present-day factories. At the moment they require workers either to feed components into a machine, or hold a component in place while the machine attaches it to the main assembly. In order to create a fully automated factory, robots will have to be developed to take over the 'handling' tasks from humans. In other words there will have to be two types of robots in future factories: 'worker' robots and 'transporter' robots.

The 'worker' robot would undertake the tasks of feeding components or raw materials into machines for assembly or tooling. Initially these robots will operate by touch alone. They will identify the shape and position of an article, after removing it from a supply in a bin or on a pallet. Then they will orientate it into the correct position for insertion into a machine. Ultimately they will be equipped with a visual sensor to perform this task quickly and simply.

The 'transporter' robot will act as a cross between a warehouseman and a fork-lift truck. It will unload goods and materials delivered to the factory and store them in their correct compartments. It will inform the central computer of existing stocks and it will also be responsible for carrying the different components and materials to their respective production machines.

Another development that we can look forward to in the factories of the future will be highly versatile machines capable of performing a wide range of mechanical operations. If we reduce the degree of specialization in individual machines it will make the altering of their functions a much simpler operation than the costly, time-consuming retooling which is necessary to convert present-day machines.

If machines are to be of use in relieving people of the monotony of working on assembly lines, they can be of similar value in replacing men in dangerous or disagreeable working conditions. It seems quite probable that in fifty years' time coal mining and other kinds of mineral extraction will be fully automated, especially since one of the major costs in mining today is that of putting miners underground in relatively safe and tolerable working conditions.

Other disagreeable types of work could be done away with by a simple change in public attitude. One of these will be dustbin collection. In fifty years' time our domestic refuse will not be carried away once a week by half a dozen powerful men calling at individual homes; we shall dispose of it ourselves.

Some of it will be compressed into the combustible brickettes to which I referred earlier. The rest will be put into plastic bags for conventional dustbins. When the bag is full, we shall carry it no more than 400 metres from the house and dump it into a communal skip. These skips could well be made into attractive objects of street furniture, decorated with scenes from Boccacio or reproductions of Lowry. Then, once a week, a vehicle will come and remove the full skips, replacing them with empty ones.

Down at the shops

Although, as I have mentioned elsewhere, it will be possible to 'window-shop' by television long before 2030, people will still want to examine the goods they buy, besides which shopping is for many people an enjoyable part of their leisure activities. This being so, I am inclined to believe that in addition to computerized supermarkets, there will be a swing back to smaller shops. In the future, large shopping precincts may become less acceptable. When we look back in the year 2030, we may regard the shopping centres of the 1970s and 1980s in the same way that we view the tower blocks of the 1950s and 1960s. In 1950, people thought that tower blocks were an advance in housing. They offered high living with a view far above the noise of the crowds and the traffic. Thirty years later our attitudes have changed. Vandals put the lifts out of order, so that people living on the higher floors become prisoners; hooligans soil the stair-wells and corridors; children cannot safely go out to play. Perhaps in fifty years we shall also come to believe that large impersonal shopping precincts are examples of the new brutalism.

An increase in the number of small shops could arise from the growth of leisure. A man made redundant at the age of fifty with a lump sum of, say, £10,000 at current values, might easily decide to open a small shop. He will already be receiving an adequate pension from his former employer, so his income is assured. The shop will offer company, interest and, with good luck, a financial return as well. At the same time, shoppers will appreciate the renaissance of small shops as a GASH.

Even today, people hurry to the small baker as soon as there is a whisper that the delivery men employed by the large bakeries are threatening to go on strike. In 2030, small shops will satisfy both the shopkeeper and the shopper whom he serves. It it noteworthy that in a country like France, which has maintained a tradition of small shops and markets, many people still prefer to buy from these than from the larger and more efficient supermarkets.

Supermarkets themselves will remain much as they are today. They are efficient, hygienic and inexpensive, and I do not anticipate that we shall be buying food in larger and larger quantities to store in a deep-freeze. Rather, we shall be buying large quantities of basic foodstuffs to store in the larder as another of our domestic GASHs. People will bring home 50-kilo sacks of flour and peas to protect themselves against a breakdown in supplies. When there is a heavy fall of snow and no one to clear the roads, the telephone lines are down and the trains have stopped running, we shall go to the larder and live off our basic rations of flour, peas and six gross of tinned sardines until things return to normal.

Of course there will be changes in future supermarkets. The major modification will be in the check-out procedure. There are many financial predictions that claim we shall soon be living in a cashless society. All our purchases will be paid for by credit cards fed into retail computers in the shops. These will instantly contact the computers in our banks to debit our accounts automatically. Such a system would avoid the risks attached to carrying sums of untraceable currency and save fumbling for change in the supermarket queue, or hold up other shoppers by writing out cheques.

In addition to receiving payment, every cash-register of the future will be able to read the magnetic price-tags on each item, firstly to record the price and secondly to register the sale with the central stock-taking computer. This will mean that a continuous stock-take will be in progress during the trading hours, monitoring the depletion of every item and ordering replacements before stocks run out. Sales personnel will also have access to up-to-the-minute information on the demand for and popularity of every line sold in the shop. As the information passes from the check-out to the stock-taking computer, an instruction will be passed to the restocking unit which will automatically replenish the shelves as they become emptied. At night when the shops are closed, the central computers will get in touch with Head Office to give the day's sales figures and receive any instructions on changes in price. Bank books will also be balanced in the same way so that the clerks will not have to stay late to do the work themselves.

Though credit cards will make a major difference to the way we pay for goods, I cannot help feeling that cash will survive for fifty years and probably for a good deal longer, but only for small, everyday purchases like matches or newspapers. Another reason why cash will retain its popularity is because of its usefulness in exchange for extra work during leisure hours. It is customary in 1980 to deprecate such unauthorized activity, but by 2030 its economic and social usefulness may be recognized. By then, I foresee work and leisure being taken out of their separate categories and blended into the all-enveloping fabric of life. Work will become pleasure and pleasure will become work, and money will be scattered over the two.

All Work And No Play

Lessons in leisure

'Increased means and increased leisure are the two civilizers of man,' said Benjamin Disraeli. Faced with the prospects of increased prosperity, an enlargement in the work force and the wide-scale introduction of technology into the whole field of employment, we may indeed hope to be more civilized in fifty years' time.

The distinguished biologist C. H. Waddington* drew a parallel between our prehistoric, hunting predecessors and our post-industrial offspring. Since the invention of agriculture, he said, people have spent most of their time working – work, by definition, being something they would not choose to do. In prehistoric times, when man hunted his food, this was not so, and it would not be so in the twenty-first century.

Just as prehistoric man (as opposed to woman) spent much of his time sitting around, chatting with his chums, killing time between killing mammoths, so working people in the future may be faced with the same prospect of choosing how to dispose of their spare, non-working hours.

Waddington developed his argument by indicating the reduction in work time that has already taken place in industrial societies during the period since the beginning of the Industrial Revolution. The average working life has been shortened by legislation, prohibiting young children from starting work before a certain age and pensioning old men and women when they attain another age. At the same time, legislation has limited the number of hours in each day that may be spent working. He points out, however, that many workers have opted to undertake additional work, either officially or by 'moonlighting', implying that they prefer 'greater affluence to increased leisure'.

Will people want to get richer and richer as the wealth of the industrialized communities to which they belong grows,

* *The Man-Made Future*, p. 215, C. H. Waddington, Croom Helm, 1978.

or will the pendulum swing away from an interest in wages to one in leisure? J. Martin and A. Norman* have estimated that if the GNP of industrialized societies maintained a growth rate of 3½%, by the turn of the century the workers in those countries would be three and a half times as wealthy as their predecessors in the 1970s. In addition, they would be faced with six months' holiday a year or, alternatively, with three eight-hour working days a week plus ten weeks' annual holiday.

Waddington discussed some of the problems that would arise from such an increase in prosperity. He foresaw the increase in manufacture using up basic natural resources and increasingly contaminating the environment. On the other hand, if, as is likely, population numbers in Western society fall as families restrict the number of their children, fewer goods would be needed to satisfy their demands. In Waddington's view there is only a limited requirement for washing machines, cars and television sets. Hence, if affluence proves no longer capable of providing satisfaction, leisure becomes the alternative way to improve the quality of life.

This trend is already apparent in some industrialized societies, even where the growth in GNP has not reached the level of 3½%. Dennis Gabor referred to the 'spreading voluntary absenteeism', which is estimated to cost the British economy from 200- to 300 million working days each year, which is one-third more than the total working time lost in strikes.

If the trend towards shorter working time and longer periods of leisure continues, in fifty years weekends will extend over three days to include Monday, the weekday with the highest absentee rate, and work will start at ten o'clock in the morning and end at four in the afternoon. Assuming an hour off for lunch, we shall see work being done for five hours, four days a week, giving a twenty-hour working week, extending over a forty-five or forty-six week year, this reduced by a six-week holiday.

Is this an attractive prospect, or are we converting a

* *The Computerized Society*, J. Martin and A. Norman, Prentice Hall, 1970.

newly leisured class into the new unemployed? Leisure in the future must be more than amusement, it must also be polite, creative, stimulating and productive as well.

Fun and games

Before reflecting on what we shall be doing in our leisure hours fifty years from now, let us spend a few moments looking at some of the wilder proposals about what I believe we shall *not* be doing.

The science-fiction writer Isaac Asimov once wrote an article entitled 'Future Fun'*. In this he set out to examine what recreation might be like in the future. Leaving aside existing sources of entertainment likely to continue into this future, he concentrated his attention on new recreations. For example, he visualized 'Hilton-Antarctica' and 'Sheraton-Greenland' for those who wished to escape from sunny beaches to something more rigorous. For others who enjoy the sea, there could be activities besides the current use of aqualungs and surfboards as playthings. These 'aquaphiles', he suggested, could live in the sea, perhaps in floating cities. For them the sea would not be merely a place of occasional recreation, it would become their environment. Asimov envisaged children in these cities learning to swim at the same stage that their terrestrial peers learn to walk. Their hobbies, he speculated, would include skin-diving in place of fell-walking, and their pets might well be dolphins instead of dogs and ponies. Then, leaving the common world and commonsense far behind, he suggested that one day people might learn to swim in enclosed areas of the ocean as water-breathing creatures, free from the encumbrance of oxygen cylinders and breathing masks. A system of oxygenating water under pressure, he claimed, has been successfully used to enable dogs to breathe under water. It is not the

* 'Future Fun', Isaac Asimov, in *Lithopinion 6*. Local One, Amalgamated Lithographers of America, 1967.

water that causes us to drown, he pointed out, but the lack of oxygen.

His next proposal was for holidays on the moon. Had not Alan Shepard, the American astronaut, already driven a golf ball there? The moon's landscape is interesting, the view of the sky is good, and the earth is worth looking at. The disadvantage of having to spend much of the proposed lunar holiday underground to avoid being hit by falling meteors and burnt by ultraviolet radiation is glossed over and there is talk of devising new rules for games and sports and of the 'mass appeal' of lunar mountaineering, which is described as 'safer and easier' than ordinary climbing, and of skiing. It is all splendid reading. No serious attention is paid to the real problems, the real dangers, the real achievements, or to the gigantic cost.

Extending his imagination beyond the time when the moon might perhaps become a little 'spoiled', Asimov guides the future holidaymaker to the ultimate resort, the orbiting space station where no gravity exists. Here, those who relished the ultimate in expensive ostentation could savour the sensations of weightlessness.

Back on earth, Asimov predicts a development of other and less extravagant ways of using technology as a plaything. He anticipates a time 'when computers will play chess at great speed and men will watch large reproductions of the swiftly changing patterns on chessboards with interest and absorption.' While crowds could watch such master computer chess tournaments, the few mathematicians who wrote the programmes for the computers could also enjoy doing so.

Most gruesome of all, Asimov envisages the development of a form of recreation taking place within the brain itself. The controlled use of psychedelic drugs, he professes, would allow men to indulge their chosen fantasies without the need to move from their own sitting-rooms. Hallucinations would provide complex recreation in a capsule. Swallow it, sit back and allow your brain to transport you to the boundaries of your wildest dreams. In an age with a superfluity of workers, some may opt to spend their entire lives in one selected fantasy world. Aldous Huxley thought about this fifty years ago and, in his parable *Brave New World*, showed it for the ultimate prostitution it is.

Already people are becoming disillusioned, not only with the vacuousness of 'pleasure', but with the rigours of flying to the Mediterranean and having to spend hours or days waiting in airport lounges. By 2030 the novelty will have worn thin and the true pleasure of being involved in creative activity, whether it is writing or digging, during the weeks in which we do not have to work for somebody else, will have become apparent.

I have referred earlier to what I anticipate will be a growing interest in building our homes. In the future, as in the present, the more resourceful among us will find real satisfaction in adding a new wing to the house, or building another storey during the holidays. Less ambitious people could take up temporary jobs in a totally different field of employment. Bank clerks may choose to become fishermen, farm workers may turn their hands to being park attendants. The man who would paint pictures if he could can beautify his neighbourhood and earn himself £5 a time by painting his neighbours' dustbins with floral designs, so that they ornament rather than disfigure the street scene.

In the future, a more rational attitude towards work and leisure will not simply be restricted to holiday times. People will employ their increased non-working time equally constructively. I see the development of such craft talents as carpentry or pottery into profitable cottage industries to become the suppliers of hand-made goods in the future.

Many craftsmen working on factory production lines will develop their skills while there will be others with the spare money to buy their products from them. Others will learn their handicrafts at night schools or adult education centres and put them either to creative production or use them to protect themselves from the helplessness of the social breakdowns which make the lives of citizens of an industrial state hazardous. Latin has disappeared from today's curriculum; tomorrow plumbing, electronics and constructive engineering must take its place.

One of the most rewarding of handicrafts has always been gardening. The idea of a large communal garden – virtually a farm – is an attractive one. On the other hand, the creaking fabric of modern society raises problems for such a proposal. Where a community has harmonious views, a co-operative garden has merits. In other places, the individual patch, providing recreation, food and space, has a greater attraction.

In 2030, gardening techniques may not be very different from what they are today. For a man who enjoys digging, a mechanical rotovator is self-defeating. Weeds, which have been one of the troubles of gardening, can be avoided by covering with black plastic sheeting those areas of earth which are not occupied by crops. Those who do not like the look of black plastic in the garden can cover it with a layer of ornamental pebbles.

Plastic can also be used to transform the greenhouse. In place of a fixed structure, there will be a movable plastic pressing like a transparent dish-cover, with a fan fitted into its side to blow in warm air and moisture to maintain a suitable atmosphere for the plants underneath. Facilities will be provided to introduce carbon dioxide or ethylene.

Mowing the lawn will no longer be necessary in the twenty-first century garden. By then, the grass that will be sown will not grow more than a few centimetres high. Nor will it be necessary to water the garden. Automatic irrigation, the major innovation in gardening in the next fifty years, will free us from worry and free our plants from over-zealous saturation. The technique has already been successfully applied on a commercial scale in Israel. As the use of plastic plumbing becomes standard, similar systems will be introduced into our gardens. Pipes will be laid in trenches, and will be equipped with indicators to keep a check on the moisture level of the soil. When it drops below the pre-set level, the indicator will activate a valve, which will open, allowing exactly the right amount of water to flow into the ground to restore the moisture to its correct level.

Many people will also be engaged in small-scale agriculture in their gardens. Writing on alternative technology,

Robin Clarke* has described a system of food production in one garden in the USA. This combination of horticulture and pisciculture produces large quantities of fish protein on a small area, and is independent of any external supplies other than sunlight and domestic sewage.

A small pond covered by a roof of timber and glass is fed with the overflow from a septic tank. Insect larvae are encouraged to thrive in this hot, humid shelter, feeding on the organic matter in the pond. Every week the larvae are scooped from the surface of this pond and fed to Tilapia fish which live in an adjacent pond. The fish pond is also covered with a heat-retaining plastic dome, which maintains the water temperature at 20–30°C, a temperature ideally suited to this species of fish. In the course of one summer, the fish grow to an edible size and are sold to wholesale dealers. The water from the two ponds is then pumped on to the vegetable garden as fertilizer. Fish ponds of a less sophisticated nature used to be common in Europe when the monasteries were at their height and monks required a readily available source of fish every Friday. They are still commonly found in parts of Asia and the Far East. Perhaps in fifty years they will have returned to the developed world.

With luck, there will also be a swing back to the rearing of domestic animals. At present public concern about dogs defaecating in public parks has brought about an irrational antipathy to household pets. The revolt will come when the sanitarians try to do away with cats. The modern attitude is a confused one because, running parallel with this hostility towards dogs, there is a sentimental interest in nature programmes on television which has led to an extraordinary weakness for wildlife. This may well have abated in fifty years' time when people will have come to realize that domestic animals are also a GASH. Dogs frighten away prospective housebreakers and discourage unwanted salesmen and missionaries for esoteric religions, and cats keep infestations of mice under control. A return to the practice of keeping chickens, rabbits and pigs in the garden will

* 'The Pressing Need for Alternative Technology', *Impact of Science on Society*, 23, 4 (1973).

ensure that there will be meat, poultry and eggs to eat when commercial suppliers are on strike. Goats will enable households to produce their own supplies of milk. Any inconvenience from the smell will be amply counterbalanced by the affection inevitably aroused by these intelligent animals and the fascination of making goats' milk cheese.

With more time to spare from work, the rearing of livestock will be found to be of absorbing interest. There are those today who derive pleasure from keeping pigeons which they do not even eat. How much greater will be the satisfaction of coaxing one's own chickens to better the egg-laying record of the hens next door! The food the animals would provide is only one of their virtues. They would also consume many of the kitchen scraps that are currently thrown into the dustbin. Their manure would do the garden good and the skins of the goats, rabbits and pigs could either be sold to dealers or else used at home in a craft hobby making goatskin slippers, pigskin gloves and rabbitskin mittens.

These are only a few examples of the imaginative and beneficial ways in which people might put their land to use in the future. By 'land' I am not necessarily restricting them to their own back gardens. Greater use will be made of smallholdings and allotments, of the acres of land beside the railway lines and in the central reservations of motorways, transforming untidy banks of grass and rubbish into mile upon mile of neat, cultivated market gardens.

In 2030, we shall see more locally grown produce on sale in markets and many households will be able to supply themselves with all the vegetables and fruit they need. What might begin as a domestic interest in growing a few carrots and runner-beans will blossom into an alternative, labour-intensive system of farming on a not insignificant scale, helping to curb our national demand for imported foodstuffs and at the same time providing useful, enjoyable employment.

A sight for square eyes

Looking ahead, and not so far ahead at that, nearly every sitting-room in the Western world will undergo a fundamental reorganization. The television will be moved from its box on wheels in the corner, to be hung on the wall. This metamorphosis will come about through the substitution of the conventional cathode-ray tube for a new device; the gas-discharge panel. At present the cathode-ray tube is too bulky to provide a satisfactory wide screen. Consequently the size of the television picture has been reduced to the limits we currently accept. In the future, however, gas-discharge panels will enable large screens to be hung on the wall. These may eventually occupy the whole area of one wall.

The panel will work by constructing a picture like a newspaper photograph, building it up from a series of dots. Each dot will be formed by a neon-filled cell, the size of a pinprick, activated by a small electric current which makes it glow. The degree of brightness of each cell will be controlled by varying the amount of gas and the level of current passing through it. With about 150 cells to the square centimetre, the quality of the picture will be superior to that produced by existing cathode-ray tubes. The wall screens will be less than ten centimetres thick and will contain all the circuitry needed for charging the thousands of cells. When we sit in front of these wall-sized screens in fifty years' time, the wide open spaces of the prairie and the tension of the Cup Final will no longer be compressed to fit into the confined space of the twentieth-century 'goggle box'. The sensation will be as if we were there, entering into the action ourselves. We shall watch enthralled as life-size Red Indians charge towards the stage-coach, fleeing across the wall towards the french windows, and cringe in the contoured seating of the fluidized-bed settee as the winning goal hurtles towards the cocktail cabinet.

Before fifty years are up these 'super-televisions' will possess a further innovation. The picture, besides being bigger, will be three-dimensional, and produced as a hologram. Holography was invented ten years ago; at present, holograms are little more than three-dimensional lantern slides. But the technology is moving so fast that three-dimensional television is a clear possibility, virtually a certainty in fifty years' time.

To make a hologram, a laser beam is split by a half-silvered mirror. Half the light passes *through* the mirror to shine directly on to a piece of photographic film. The other half is *reflected* by the mirror on to the object to be 'hologrammed' and then, in turn, reflected on to a film. When they are joined together on the film, the two beams of light form what is called an 'interference pattern', a pattern of dark and light shades. By beaming a laser back through the hologram slide, the original image is reconstructed in three dimensions. Once the technique of recording moving images on holograms is perfected, three-dimensional television will have arrived. Then Father will be able to sit back and watch the dancing girls, and if he should decide to take a closer look at the one in the back row he fancies, he will only have to move his head so that the comedian in the front is no longer in the way.

For those who relish the prospect of playing a more active role in their television viewing, video cassette recording will be enormously improved. Self-editing machines for use in the home will allow the action to be slowed down or replayed. It will be possible to enlarge any part of the picture into a close-up, to divide the screen into segments, or to introduce inserts from other sources. It may even be technically possible to enable viewers to participate in broadcast discussions through control panels set into their television receivers.

Night on the town

It is difficult to foresee what direction public entertainment will take in the next fifty years, but technical advance is certain. Television we have just discussed. Its technical possibilities make it capable of presenting drama, music and entertainment more directly and forcefully than has ever been possible before. This, however, is no insurance that the programmes it transmits will be worthy of the expertise that makes their transmission possible. We can today put up a satellite to circle the globe by which programmes performed in Europe can be seen in America. It is for us to choose whether to transmit *It's a Knockout* or *Hamlet*.

There is evidence for hope that the growing wealth and increasing leisure of twentieth-century life will, in the twenty-first century, lead men and women to enjoy and participate in creative and worthy entertainment. Already we have seen the growth and development of live theatres, not so much in capital cities, but in outlying towns and boroughs. Music also is flourishing, whether in the form of classical music, brass bands or pop festivals. Today, as leisure grows, such activities will grow too. There is a movement for the establishment of small companies of actors, singers, musicians and dancers to provide local entertainment in the areas in which they live. Some of the performances may lack style and polish, but the stimulation they provide is valuable nevertheless. Painting and sculpture, like music and drama, show signs of being recognized, as the fifty years pass by, as the business and delight of everyone, not merely of a specialist minority. Perhaps the 'new brutalism' in which we have lived, the concrete monsters and the sordid streets, will bring their own reaction. I have suggested that the dustbins which have come in our times to disfigure the most elegant areas of our cities should be decorated to please rather than disgust the eye. This could be a beginning followed by groups of masons volunteering to carve the blank face of the ruthless concrete environment, serving, as did the work of the medieval craftsmen who mocked their masters in effigy as gargoyles, to civilize the surroundings of those who live in 2030.

In The Know

Extra, extra, read all about it

Man has been eager for news and gossip for as long as he has lived in communities. From the earliest Acta Diurna which carried accounts of Julius Caesar's exploits and was plastered over the walls of ancient Rome, to the wall posters that denounce offending officials in present-day China, news has always been a source of fascination.

At present, most of us have to walk as far as the front door every morning to collect the daily paper. However, it is not unreasonable to predict that if certain developments in mass communications come to fruition, we may expect to walk no further than to our television sets in order to receive our morning edition of the 2030 paper.

These projected newspapers would be the end-product of microtechnological automation. From the editorial office to the breakfast table, the process would be governed by computers.

Reporters and columnists would type their articles on video-typewriters connected directly to the computer memory in the editorial office. As they type, their copy would appear on the screen in front of them. The computer would correct their spelling and punctuation before storing the article in its memory. The editor would later retrieve the articles, to check and alter them as necessary, before dispatching the copy to be typeset by a computerized phototypesetter, which would subsequently produce the finished master copy to be stored in the central memory bank. All other copies of the paper would then be copied from this one.

Instead of reading the same newspaper every day because it happened to be the one delivered by the newsagent, the advocates of computerized newspapers propose that we should have a varied choice in the future. Selection and delivery would merely be a process of tapping out a number on the keyboard of our home computer. This would put us in contact with the main computer holding the master copy of the newspaper we wish to buy.

Overseas newspapers would be transmitted by satellite and the signals would be received by a small dished aerial attached to the chimney. Those from computers nearer home would be transmitted by fibre-optic land-line. A laser printing unit attached to each home computer would print for us a copy of the newspaper we had selected within four seconds. On the other hand, we could choose, if we wanted to, to have it microfilmed or displayed, page by page, on the screen of our TV set.

It is difficult to foresee whether this easy access to instant information would lead to people reading more newspapers than they do now; whether they would stick simply to one favourite paper or choose a few pages of a specialized newspaper every day. It would, of course, be possible to arrange to receive a personalized newspaper, specially compiled to suit our individual interests and tastes. Data about our occupations, education, hobbies, interests and socio-economic level would be housed in a central scanning computer, which would thumb through the world's newspapers, periodicals, films, television programmes and radio broadcasts. Whenever it came across anything that tallied with our character profile, it would extract it and thus compile a newspaper, every page of which would be guaranteed to interest us. Such a paper might easily be made up of the front page of *La Stampa*, letters to the *People's Daily*, the sports page from the *Washington Post* and the fashion page from *Pravda* with a selection of notes on the first hearing of cuckoos from *The Daily Telegraph* and the *Whitby Gazette*.

Even though we receive only a limited amount of information today under the guise of news and hear it repeated too often – in the paper, on the radio, on the TV, every hour of the day and for most hours of the night – it is by no means certain that because in 2030 it will be technically feasible to gather for ourselves news from every quarter, we shall take advantage of the possibility. As it is, we *can* listen to radio from all round the world. In fact, few of us ever trouble to do so.

News is what journalists put into newspapers and news bulletins. Because we get news from a number of different channels, it does not follow that we are any better informed. The same old news, put out by the news agencies, is reproduced in all the newspapers. Interesting events that are not reported do not become news now. It remains to be seen whether they will fare any better in 2030.

Similarly it does not follow that because we are subjected to an almost incessant barrage of information we are, in fact, well informed, for what we hear is the limited amount of information selected as being 'news' repeated many times over. Perhaps a tyrannical ruler who decreed that there would be only one ten-minute news bulletin each day at eight o'clock in the morning would produce a better informed, more sensible and more thoughtful community by compelling journalists and editors to publish interesting facts not currently characterized as news.

People like newspapers and, with all their faults, I see them surviving to 2030 and beyond. Great is the word, says the Bible, and greater still is the printed word. A change that can be foreseen in the newspapers of 2030 is a technical one: they may no longer be made of wood. The demand for newsprint is one of the major contributors to the present destruction of the world's forests. There seem to be three ways in which by 2030 we may be able to continue to read the papers without deforesting the world.

Firstly, the size of newspapers might be reduced and the news printed in smaller type. Secondly, newspapers could be printed on washable nylon sheets, to avoid the necessity of cutting down so many trees. The third possibility, popular with those concerned to preserve the environment, would be to recycle newsprint. Technical problems in doing this have been looked into for a number of years. The principal difficulty in using recycled paper is that it is weaker than new paper. Furthermore it tends to retain traces of ink, which produces unsightly speckles on the finished page. In the future, we shall need a process to clean thoroughly the old ink from the paper. The main problem to be overcome is that of organizing the collection of used newpspapers to be recycled, but it should not be insuperable, as many groups of people do it now. In 2030, which we have agreed may be, if not exactly an age of leisure, then an age of more leisure, one of the responsibilities of those with the most free time may be to organize and oversee the collection of old newspapers.

Perhaps we shall see a combination of changes in our twenty-first-century papers: fewer, smaller, more closely printed newspapers; some printed on new paper, some on used paper, some on nylon; and all supplemented by an electronic newspaper read from the TV screen. This could also carry the nude girls which are a *raison d'être* of some modern newspapers. Once we had looked at them we could then make them disappear, thus relieving basically decent citizens of their share of having no bother with them.

The price of sausages in Mogadishoo

One of the great hopes of the future, as I have said, are home computers. Apart from providing newpspapers, these microtechnological domestic genies could be expected to regulate almost every household operation, from watering the plants to letting out the dog. These, however, would be trifling auxilliary tasks in comparison with the computer's chief function of providing virtually limitless information, and facilitating almost boundless communication.

Essentially, the home computer would act as a domestic extension of the centralized computer banks of information, or communications systems, much as televisions and telephones do today. There would be a two-way, interactive contact, with the computer both receiving and transmitting domestic messages.

Initially these home computers might resemble existing telephones, through which the user would talk to the computer, recording its replies on a video-screen. To establish contact, we would dial the number of the computer we wished to consult, which could be anywhere from Auckland to Anchorage. Once contact had been established, we would tap out our message and watch for the computer's reply on an adjacent screen. The PRESTEL system in Great Britain works in this way, using existing telephone and television equipment. In the future, the system would be extended to an international network providing the caller with information as diverse as the price of sausages in Mogadishoo and the starting prices for the three-thirty at Uttoxeter.

Later, as the microtechnological capabilities of the silicon chip unfold before our eyes, it is predicted that home computers will be able to take on increasingly complex operations. We may become the owners of machines on which we can write and draw, like those coming into use in drawing-offices, which were mentioned earlier. By drawing with a light pen, we could send pictorial messages as well as verbal ones to the central computer.

Computers in 2030 offer the possibility of a return of the home-dressmaker – more skilful, quicker, more versatile – and electronic. A lady wants a new dress. What will she do? Sitting at her home computer, she calls up the couturier, whose computer, in its turn, knows her measurements. She tells it she would like to design a dress. The computer flashes a diagram of a standard female figure on to the lady's screen. With a light pen she sketches on the female figure the kind of dress she has in mind. Only a rough sketch is needed; the central computer has fashion data in its memory and instantly draws the finished dress. The customer is not quite sure it is what she wants and asks (by pressing the appropriate buttons) the computer to show her some other patterns. It does so. The customer chooses one. At her request, the fashion-house computer flashes up on the screen a diagram of her wearing the dress. She sketches with the light pen a few modifications which the computer draws properly. She asks it to redraw the dress made from a different fabric. It does this too. She asks it to cost the dress. Up comes the price. She decides to take it, but first she asks for a close-up of the collar. No sooner said than done. So she tells the computer her credit-card number. It makes a note of it, verifies that her credit is good, debits her account with the cost of the dress, and sets the machinery in the factory in motion to assemble the parts and put the dress together. Within an hour, it is made exactly to the customer's recorded measurements. Before she finishes the transaction she presses the button to tell the computer where to deliver the dress – to her home, or to a convenient pick-up point. And that's the world of 2030.

Popping round the corner to buy a paper or a special rye-bread cake is one kind of shopping. We have touched on this already. Stocking up the week's groceries is another. Here's what it will be like in 2030 or, probably, well before. First the customer will get her computerized TV in touch with the supermarket and ask its computer what special offers are on sale. She will sit in the lounge watching the articles flash by on the TV screen. Then she'll get down to business. Working through her shopping list, she taps out the code number for each commodity she wants and tells the computer the amount. When she has finished she asks the supermarket computer to play back the list so that she can check it. It does so. She has forgotten something; she adds it on. She made a mistake before; it corrects it. Right. She finalizes the order, tells the machine how to get the goods to her, tells it her credit card number, and that's it. Her own machine will give her a punched tape as a record of the transaction. Of course, while all this has been going on, at the supermarket packets of cornflakes and boxes of eggs have been sliding down chutes in obedience to the computer's instructions as the lady's order is assembled. When she made a mistake, the biscuits she first ordered had to be shunted into a siding and the bacon slid down in its place. And all the time, the supermarket stock records were clicking out the items.

Electronic 'window-shopping' can provide a great deal of information. It shows what is available and at what price. But it cannot feel the apples the way the prudent shopper does.

The British Post Office has developed the system they

call PRESTEL which is the beginning of the instant information services which will be commonplace in 2030. The PRESTEL station broadcasts a wide range of information in addition to normal television programmes, using the same signals for both. Home television sets capable of recording this information are fitted with a special device for extracting, decoding and storing the transmitted information. By using a push-button selector, the viewer can select the information he wishes to see. This will be displayed either on a blank screen or superimposed over whatever programme is showing at the time. Using this system, one can receive the latest weather forecast – from Manila if you like – without having to sit up half the night to tune into Radio Philippines. One merely refers to the directory provided, taps out the number for world weather forecasts and goes to bed. In the morning, one can retrieve the stored information which shows up on the TV screen. The system can provide a much wider range of information than this, however.

Speaking with light

One of the major innovations of the next fifty years will undoubtedly be the introduction of fibre-optics into communications. These have already been used on a limited scale for several years, but only in circumstances that have permitted them to be installed in a project, for example a new town, from scratch. Their major potential value will be in transmitting television. Up until now both radio and television broadcasts have been limited by the radio frequency spectrum. In the future, however, at least some of our television programmes will be transmitted through fibre-optic cables. Houses will be 'wired' with cables carrying both television and radio transmissions.

Instead of the television companies being obliged to transmit on only a few channels for fear of interfering with each other, they will be able to offer a variety of programmes designed to appeal to specialist or minority groups as well as the large numbers who only want to be entertained. Local television stations will be able to provide much more detailed coverage of local events. The quality of the picture will be far better, because cable television is not subject to atmospheric interference.

Fibre-optics will also play an even more important role in general communications. They will enable the gas and electricity companies to read domestic meters from a central scanning point and avoid the necessity of employing meter-readers.

In fifty years fibre-optic cables will have replaced existing telephone wires, because they are very much more efficient. Light is an electro-magnetic radiation of a particular wave-length. As such it can be used to work telephones or transmit TV. As laser light, beams can, it has been found, pass with virtually no resistance within a fine glass 'hair'. This is what fibre-optics means. The great advantage of fibre-optic cables lies in the enormously greater capacity of light to transmit information than other forms of electro-magnetic radiation. This is because light waves have a much higher frequency than radio waves. The light pulses from the laser can carry about 4000 telephone conversations simultaneously through a fibre no thicker than a human hair. Cables made up of 100 such fibres would be about 7 millimetres in diameter and would allow 400,000 telephone calls to be carried in a cable no thicker than a pencil. Lasers the size of pin-heads can be used to transmit more than 120 million pieces of data every minute through such cables.

The potential advantages to be derived from fibre-optics are enormous. The installation of the cables, however, will be expensive. It requires a large national investment programme if every house in the country is to be connected to a central 'exchange'. Of course, in new towns, the installation of the cables is no more difficult than installing mains electricity and water services.

A word in your ear

Changes in our capacity to communicate may well be one of the most significant differences between 2030 and now. Fifty years from now letters and telephones will still exist in one form or another. The development of systems employing home computers will, however, open the new possibility of being in contact with any part of the world through home computer exchanges. Videophones, which enable people both to see and hear at a distance, already exist and the complementary development of fibre-optics will greatly facilitate their rapid dissemination.

There is reasonable cause to anticipate that we shall be using portable telephones well before 2030. Devices of this sort already exist and in fifty years it is very likely that most of us will be walking about with telephone walkie-talkies, just as we carry briefcases and bags today. The caller will tap the number he wishes to call on a push-button console and will then speak into the mouthpiece. His conversation will be carried by radio signal to a central monitoring computer from whence it will be transmitted by land-line or satellite to its destination in any part of the world. It might even be possible to dial any number in the world directly through this system. As a refinement, it would be possible to record on miniature magnetic tape cartridges numbers that were frequently required. These cartridges would then 'dial' the appropriate number automatically when they were slotted into the portable phone.

A variant of a portable telephone, but lacking its privacy, is the so-called Citizens' Band Radio, which has become popular in the USA and Australia. This is not an innovation for the future, it is a commonplace for the present, being used by taxis, the police, the fire service and many others. So far the authorities will not permit its use in Great Britain.

Reading, writing and arithmetic?

The kind of education which will be considered appropriate to the society of 2030 will depend on what kind of society we want to have then. Looking back into the past, we see the foundation of our great and ancient schools by devout princes of the Church anxious to give to a few poor boys the precious knowledge of reading and writing. The universities of the 1930s, although having long included science and engineering within their syllabuses, were still based on the classical tradition. Our own utilitarian generation has thrown out as a waste of time the attention given to Latin and Greek, to the study of grammar and the memorizing of verses from the poems of the ancients. Now that we have had time to assess what kind of a world we have produced by our attention to such 'practical' topics as technological economics, nuclear physics and brain surgery, we may find it useful to consider the reasons underlying the educational targets of our predecessors.

The purpose of the classical education of our ancestors was to allow them to read and consider the philosophy and poetry of the golden age of civilized man in that short period of human achievement – as it was understood – in Greece and Rome. This was the time when man was at his most heroic, when great deeds were done and great poems written. The thoughts of the philosophers, the deeds of the heroes and the songs of the poets could – or so it was thought – serve for the emulation and education of the civilized societies of Europe.

A parallel thread in the education of our grandparents underlay the efforts of philanthropic people to provide universal education so that men and women could read for themselves the words, first of the Bible and, later on, of those whose minds had enriched mankind. Although there are now people who indict mass education as a mere means to produce a literate workforce, those whose altruistic efforts brought it about knew that they were giving their poorer fellow citizens a rich gift. Before education was available to them, they were unfortunate indeed:

Knowledge to their eyes her ample page,
Rich with the spoils of time, did ne'er unroll.
Chill penury repressed their noble rage
And froze the genial current of the soul.

Today, ideas about our souls have changed. We have considered it a waste of time to compel boys and girls to spend laborious hours learning 'dead' languages when they could do something more useful. Our souls can appropriately expand through the understanding of science which, almost within the last fifty years, has overcome tuberculosis, poliomyelitis and virtually all the infectious diseases of the past, and revealed much of the very nature of the universe. But the basic change in our own outlook on education has come from our scepticism about the virtues and heroism of the great men of the past, whom our predecessors thought it axiomatic to admire. Our own great gospel of the equality of man prevents our claiming a higher status for the man who spends arduous nights studying in his library than for he whose nights are spent applauding a stripper in his works social club.

The potency of science and of the technology built on it have become so obvious in our present time that there is good reason for us to plan the education we provide so that we shall supply useful knowledge. To insure an appropriate supply of teachers of mathematics and electronics, of chemistry and biology, is obviously sensible to enable us to keep pace with our industrial competitors, the Japanese and the Germans, the Dutch and the Americans. In recent years, technology itself has been used to teach technology and other useful subjects, for example, business German. Appropriately programmed computerized teaching machines can be produced, into which individual pupils can be plugged and from which they can be taught any topic they choose, skilfully, accurately and at a rate specifically regulated to their ability.

Is this the model of teaching in 2030? I doubt it. Machines can be useful for particular purposes, just as books (which are the product of an earlier technology) are too. Books have played a part in the schools over the ages as teaching machines will also do in 2030, but neither one nor the other can take the place of a teacher. There is no magic in machines. They ask the pupil a question, his answer is wrong. The machine says so and rephrases the question. The pupil's answer is half right. The machine tells him about it, and so the dialogue continues. The effectiveness of the whole operation depends on the knowledge and judgement of the man who worked out the programme put into the machine at the start. How true the statement is – for books (not this one, of course!) as for the computer in the heart of the teaching machine – 'garbage in, garbage out'.

Television is a further technical aid for education in the present, as it will be for that of 2030. Undoubtedly it has its value. It allows pupils to 'meet' the leading instructors of their times in the same way as jet aircraft also serve as educational tools by enabling school parties actually to visit the pyramids on the Nile. Those of us who have returned from package tours to Turkey, let us say, are well aware that we have widened our experience although it is doubtful by how much our education has been extended.

A paradox of our times is that even while we accept as axiomatic that education must be 'useful', there is an increasing pressure for it to be useless. This was shown by the swarms of university students pursuing 'relevance' and taking degrees in sociology rather than in accountancy. And in the schools, truancy is widespread, mathematics unpopular and the reaction against tough teaching, hard learning and testing examinations overwhelming. This has come about because of an eminently desirable drive to expand the personality and emphasize the worth of each individual child, regardless of what that worth is.

So what of the future? What kind of a place will the school of 2030 be? We shall in fifty years, as we do now, complain about our economic troubles. The age of the silicon chip, however, will be an age of leisure. So-called 'unemployment' will be an accepted feature of our society of 2030. Our educational systems must, therefore, be an amalgam of 'useful' subjects calculated to support our technological inventiveness and 'humane' subjects permitting us to live our 'leisured' life to the full. One major change in emphasis I foresee. Today there is some contempt for history. Are we not, we imply, doing scientific things like going to the moon, so new that what our ancestors did long ago can be of little value to us? We forget that precious few of us *do* go to the moon. Most of us are faced with the stadard dilemmas of life, most of us have to choose whether to drift with the current or stand by our beliefs like honourable men. These are the very problems with which people have been faced throughout history. I foresee that in 2030 we shall recognize this and devote more time in our schools than we do now to history. As Edward Gibbon wrote, 'History is a register of the crimes and follies and misfortunes of mankind', which we would be wise to avoid, would we not?

Going Places

Horseless carriages

We have witnessed in the last half century the very high esteem that the people of the developed world have attached to the freedom of individual transport. In spite of the recent price rises in the cost of fuel, this has remained true today and it will certainly hold true for the next half century. However, conscious of dwindling oil supplies and always mindful of environmental pollution, future gazers at the passing scene have determined that private transport must change.

The present 'bogeyman' of the transportation world is the large American motorcar, the so-called 'gas guzzler'. Today 95% of the travel in urban areas of the USA is undertaken in cars: 80% of American households own at least one car; 30% own two and 10% three or more. More than 40% of the total energy consumed by the individual American is accounted for in running his car. Great efforts are being made to develop motive forces that improve on the relatively low efficiency of the internal combustion engine, which at its best performance achieves only 20–25% efficiency, and to find cars that run on fuels other than expensive hydrocarbons.

The Stirling engine, patented back in 1817, is one of these bright hopes for the future. It is an external combustion engine. It burns its fuel externally to generate heat which heats a working gas (helium or hydrogen) inside the cylinders. The compressed gas expands with the heat, pushing the pistons down. As the machine completes its cycle, the pistons move up the cylinder to compress the gas once more. This engine can run on almost any type of fuel. It is more efficient and less noisy that the internal combustion engine and it causes less pollution.

It is easy to forget, however, that the internal combustion engine itself replaced an earlier but much less efficient form of car engine, the steam engine. Indeed Nicolas Cugnot, the inventor of the first steam-powered vehicle, became the first motorist to be imprisoned for a driving offence, when he accidentally drove his motorized gun-carriage through a wall at a speed of five kilometres an hour. In 1906, a Stanley steam car, travelling at just under 200 kilometres an hour, broke the world land-speed record.

Like the Stirling engine, the steam engine is an external combustion engine. There is no need for the fuel to be burned under compression in the cylinder, which implies that it can burn at a lower temperature and emit fewer pollutants. The heat it generates is passed to a boiler which forms steam and this is used to drive the pistons in the engine. The new generation of steam engines pioneered by the Lears Motor Corporation in the USA might become a model for future means of powering cars if their efficiency could be improved to equal that of internal combustion engines. They work on the principle of an enclosed steam turbine. A specially developed fluid is used in place of water because it does not freeze in cold weather and causes less wear to moving parts. The vapourized fluid passes through a throttle control, turning a turbine wheel, and then passes into a condenser where it returns to its liquid state to be recycled by a pump.

One of the significant advantages of steam engines over existing car engines is that they are able to pull better at low speeds. This means that a steam engine fitted into a car would only require between half and two-thirds of the power needed by its petrol-burning equivalent.

The gas-turbine car is another project from earlier days of motoring history that there is some possibility of resurrecting in the future. The Rover Company in Great Britain launched the first working model in 1950 as JET 1, a name it lived up to. In 1963 a gas-turbine-driven Rover-BRM covered the 4108 kilometres of the Le Mans twenty-four-hour race at an average speed of 176·54 kilometres per hour. Two years later, another turbine-driven car came tenth and in 1967 a Paxton Turbocar won the Indianapolis 500. Gas-turbines seem to offer the advantages of reliability, durability, lightness and low maintenance requirements. Their major drawback is their high fuel cost. Which effectively brings us back to square one, for the foreseeable future at least. It is, unfortunately, hard to equal the mechanical efficiency of a diesel-powered internal combustion engine.

Breaking right away from the self-propelled vehicle is the ingenious conception of the 'magnetic motorway', envisaged by Professor Thring.* Here a 'magnetic field device' placed underneath a motorway would propel special carriages at 120 kilometres an hour between acceleration and deceleration branch points. Travellers would be required to hire a motorway carriage, which would travel under its own battery power to the nearest motorway branch point. There the electromagnet fitted to the carriage would take over and would accelerate it to the uniform speed of the traffic stream, slotting into the first gap in the stream of evenly spaced vehicles. The traveller could then sit back to read his newspaper in comfort, or admire the scenery, until the carriage arrived at his pre-set deceleration point, where it would automatically branch off the motorway, the batteries would take over once again and the traveller would drive to his destination.

Appealing as these ideas are, we have to remember, however, that the internal combustion engine has been with us for a long time, and, as I have pointed out, is a reasonably efficient converter of fuel into work. The motor industry has invested millions of pounds, dollars, marks and yens in its development. Some quite remarkable revolution would have to take place in the next fifty years for it to alter drastically its commitment to this traditional form of power.

Alternative fuels have been suggested as likely successors to oil. There is a strong lobby in favour of methanol, a liquid fuel, also called wood alcohol. This clean-burning fluid can be produced from vegetation, from rubbish, from coal or from hydrocarbons. Before the introduction of paraffin, methanol was used as a lighting fluid, illuminating the streets of no less a city than Paris in the middle of the last century. During both world wars, thousands of cars were

* *Man, Machines and Tomorrow*, M. W. Thring, Routledge and Kegan Paul, 1973.

converted to run on this fuel, which has a higher octane rating than petrol. Probably its greatest significance for the future is the fact that 15% of methanol can be added to normal petrol without any need to modify existing petrol-burning engines. Ethanol, the familiar alcohol we drink, could also be used as motor fuel. It can be produced from any available source of carbohydrates. It has already been suggested that in wet tropical areas starchy root crops, of which enormous yields can be obtained, should be grown specially as the raw material for alcohol production. By this means, currently poor countries might become, if not oil-rich, at least producers of useful supplies of energy. But in the main motor cars are going to run on petrol, diesel, or bottled gas in the foreseeable future.

However, it is possible that in the next fifty years we shall be driving vehicles run on hydrogen, the use of which has many attractions. It is clean, non-polluting – the only by-product being water – and we can make it with our nuclear power. Although almost any type of engine can be run on hydrogen, its principal disadvantage as a fuel for motor cars is that its characteristics make it difficult to liquidize and consequently bulky to carry about. Scientists are currently examining the possibility of producing a metal-hydride, a sort of hydrogen-saturated metal 'sponge'. This concept is based on the fact that pure hydrogen can be absorbed in a condensed layer on the surface of certain metals, of which aluminium, iron and magnesium are the most common. If this system is adopted in the future, refuelling will be simply a case of exchanging one expended fuel 'cartridge' for a fresh one.

There is another system undergoing tests which employs a catalyst to create hydrogen from petrol or paraffin. Existing petrol engines can be modified to incorporate alternative equipment to replace the carburettor and the inlet manifold. The air-fuel mixture would be drawn into the engine by the action of the cylinders, as in current petrol engines. This would remove the problem of having to deal with hydrogen under pressure. Quite apart from the advantages of lower pollution, these modified hydrogen-burning engines have been estimated to be 30% more efficient than present-day ones.

As an admirer of Rudolf Diesel, I foresee a great development in the engine that bears his name. This operates by burning a cheaper, heavier oil than the more refined oil used in petrol engines. It also operates without the need for a special ignition system. It burns its fuel at controlled pressure and temperature during the 'power stroke' and the fuel ignites spontaneously. In these relatively stable conditions, the diesel engine converts about 35% of its latent energy into power, which is about 10% more than even the most efficiently tuned petrol engines.

Runabouts

As well as the main family vehicle, many households will own or wish to own a second car for short runs down to the shops or up to the golf course. Ideas on these range from hydrogen-powered hang-gliders to electrified bubble-cars.

In the early days of motoring, electric cars competed strongly with petrol and steam vehicles to capture the market in private transport. Five hundred and sixty-five different makes of electric car were built between 1896 and 1939. In fact, in 1899 an electrically-powered vehicle became the first car to travel at over one mile a minute. Electric buses were running in London at the end of the last century and many people have had their milk delivered by electric milk floats for more than twenty years.

The electric vehicles of the future will probably look like compact, motorized, two-seater jam-jars. Silent, cheap, clean and reliable, they will no longer be powered by the bulky lead-acid batteries that have so hampered their design until now; they will be powered by fuel cells instead.

These fuel cells will work on the principle of utilizing the reaction between sodium and water. When these two elements come into contact, a violent reaction occurs and from it emerge hydrogen, sodium hydroxide and a lot of heat. In the fuel cells of the future, this reaction will be controlled to convert the heat into electricity. The released hydrogen will also be available as a source of extra fuel. The output of such a unit is estimated to be of the order of one hundred times the output of an existing lead-acid battery of the same size. The only attention that would have to be paid to this power source would be the adding of the two basic elements, water and sodium, as they were consumed. Otherwise the cell would continue to produce hydrogen and electricity indefinitely with no risk of it running down if its owner forgot to charge it up at night. A problem facing those working on the introduction of this type is that sodium does not exist as such in nature and considerable energy is required to reduce it to its elemental state. It is also dangerous stuff to carry about.

Another projected idea of electric car propulsion is the use of a flywheel. A flywheel is a means of accumulating momentum. Momentum has been employed to propel vehicles, for example, in explosives factories where internal combustion propulsion would create a fire hazard. Little use so far has been made of it for passenger vehicles. If it proves feasible, the following could be what would happen.

Imagine that you want to pop down to the shops to buy a few groceries in fifty years' time. A quarter of an hour before you want to leave, you will switch on your little momentum trolley-car that is plugged into the mains in the garage. This will start to spin a heavy flywheel inside the vehicle. When you are ready to leave, you will open the garage doors, disconnect the power lead from the mains socket, climb into the perspex cab, close the roof, let off the brake and go bowling off to the High Street, silently propelled by the almost frictionless flywheel spinnning behind you. Supposing you get delayed in a traffic jam or you find that you have to go further than you intended because the shop has sold out of soya-bean sausages, you naturally run the risk of the flywheel gradually slowing down and coming to a halt. Fear not. All you will have to do will be to manoeuvre the trolley-car into a lay-by or parking space and plug the power lead into a socket attached to a parking meter. Even if there is no meter available, all is not lost, you will simply have to provide your own power: thigh power. You will put on the brake, open the roof, climb out and go round the back of the trolley-car to open the boot. Inside there will be a collapsible bicycle saddle and pedals, with a drive attached through appropriate gearing to the flywheel. You will unfold the seat, climb on and start pedalling. The heavy wheel will gradually revolve faster and faster until you have built up enough momentum in it to get you home again. When the wheel is spinning fast enough once more, you will collapse the saddle and pedals, close the boot, climb back into the cockpit and continue your journey. Of course this little trolley-car will not take you on long journeys, unless you are prepared to stop and build up power either from an electricity supply or by using thigh power every twenty miles or so. But it could be an effective method of getting about the neighbourhood, visiting friends or taking the children to school.

Bicycle power will once again become important in its own right. The human thigh is a far more efficient machine than many of the combustion engines in use today. The bicycle is probably the highest level of perfection achieved by technology. I am not, of course, suggesting that our entire transport system can be replaced by bicycles, but in selected areas bicycles would be the ideal means of travelling. We might see the development of bicycle-powered vehicles capable of carrying several passengers, miniature versions of the cyclomotives described on page 147. These would provide pollution-free transport while avoiding the financial pressures of ever-rising fuel prices and motor vehicle taxation. Unlike those carried by rickshaws in the Far east, all the passengers in these cyclomotives would be expected to pedal. Exceptions would, of course, be made for the elderly and infirm.

A multi-purpose pedal-powered vehicle designed in Sweden primarily for use in the Third World could be used to advantage in the developed countries in fifty years' time, if not sooner. A combination of a tricycle and a large wheelbarrow, this machine is capable of transporting heavy loads on metalled roads and even over rough terrain. In urban areas it would be a very handy runabout for family shopping or visits to the allotment.

Better safe than sorry

Road-safety is a 'crusade' which we can expect to continue for the next fifty years. One idea which has received favour is that of removing the human element from driving altogether and letting an automatic chauffeur do the job instead. Predictions have been made that the general introduction of automatically controlled cars would cut road accidents by 40%, while allowing 50% more traffic on to the road network.

The proposed system would be based on the same principle as that of the magnetized motorway. An electric current passing through a buried cable would create a magnetic field. This would induce a voltage into the electric sensors of the car travelling above. The current would then be passed to an automatic steering mechanism which would control the front wheels. The car would be automatically slowed at junctions and in traffic, and could be accelerated again when the road was clear.

In both automatic and human-driven cars, it is proposed that in future front-mounted radar will be used to track the position of the vehicle in front. Bouncing off reflectors carried on the rear bumper of the car ahead, radar signals will inform the motorist or the automatic chauffeur how far his car is from the one in front, warning him when he gets too close. A similar system using ultrasonic waves would also be able to indicate to the driver the speed and position of vehicles moving up behind him. This would prevent him from pulling out into the path of an overtaking vehicle.

For those unfortunate enough to become involved in a road accident in 2030, there will be an array of safety devices to minimize the risk of injury. Impact-absorbing cushions of water or crushable polystyrene balls might be attached to the bumpers, the motorway crash barriers and to bridge supports. Inside the car, safety belts will lock, the steering-column will collapse, the ignition will be automatically switched off and the fuel lines will be drained into safety reservoirs. The bodywork will fold up further to absorb the force of the impact and the engine mountings will break causing the engine and gearbox assembly to slip down to the ground away from the passengers inside the car.

It is my opinion that mechanical improvements will continue up to 2030, but that the pursuit of total safety will be abandoned. 'Call no man happy till the day of his death, he is at best but fortunate,' said Solon, the Athenian lawgiver. The only way we can be sure of being absolutely safe in a motor car is not to ride in it.

Naturally, we want reasonable road safety, but this has got to be a rational balance and the rationality will change as public opinion swings in the future. Speeds may well remain at the levels they are today. The technical improvement in the performance of vehicles may well be offset by a growing stringency in fuel cost: the usage of fuel will always increase at a faster rate than the increase in speed.

Changing attitudes to safety, speed and cost are social rather than technical matters. The thirty-mile-per-hour speed limit in built-up areas of Great Britain was brought into force in 1935. The technical constraints were obviously different forty-five years later in 1980 and will be different again ninety-five years later in 2030.

The long struggle to introduce the compulsory wearing of seat-belts also shows the interaction of reason and feeling. Seat belts undoubtedly reduce road casualties, yet people are only prepared to accept what they feel to be a reasonable amount of safety. There is equally good statistical evidence to ban boxing, Rugby football and ice hockey, but in 1980 and, it seems likely, in 2030, the balance of opinion favoured the freedom to be injured above the probability of harm.

As cars get faster they will get stronger and safer. The advancement in the technology of tyres is a dramatic example of this fact. You hardly ever have a puncture nowadays and if you do your tyre does not collapse and come away from the rim. Technological advances like this will continue. Safety and freedom from pollution form one side of the equation; weight, cost, complexity and inconvenience are the other. Today's average car weighs ten times as much as its driver. Hence, 10% of its energy is used to move the driver and 90% to move the car. By 2030, shall we want to pay to move more driver or more, safer, slower and heavier car?

Two driving aids that I have not so far mentioned will certainly be in use by 2030. The first is a device to warm the driver when the road is icy. This will be particularly beneficial in a country like Britain where the winter weather is variable. A second valuable innovation will be a lamp that will enable a motorist to see in fog. It could incorporate either laser beams or echo sounding equipment, and is so urgently needed that it must surely lead to its own invention.

All together now

There are those who argue that people will have to be weaned away from their dependence on private transport in the future. They point out that in urban areas motor cars pollute the atmosphere, clog the streets with traffic jams when they are in motion and when parked by the road-side are generally a nuisance. Is the solution public transport?

Ideas have been put forward for a system based on constantly moving conveyor belts. Today's underground railway relies on individual trains, between each of which gaps must be left to avoid collisions. Postulated conveyor-belt systems would carry passengers on their entire length, thereby transporting three times the number that is possible by train. Journeys on these future networks would be quicker, since the system would never stop running. There would be no need for the belt to stop to let passengers off. The passenger would step off the moving belt at his destination, either on to the stationary platform or on to one, or a succession, of slower belts.

Another way to get on and off would be to use a special lift to get up to the overhead track along which the belt would be travelling. As the lift went up, it would gradually start to revolve so that at the disembarking point it would be revolving slowly at the same speed as a large circular floor. The lift itself, when right up, would form the centre of this circle. The passengers would step out on to it, feeling no sense of acceleration since they would be revolving at the same speed. In order to join the travelling belt, they would walk to the edge of the revolving circular floor, where they would be accelerated, by the greater speed of the outside edge, to the same speed as the belt against which the edge of the circular floor would abut. The passenger would then step on.

This system would replace the noise, dirt and discomfort of our present-day subway systems with the low hum of the conveyor-belt. It would be to trains what escalators are to lifts. But the belts ought to be better than trains – there would be more of them, a train, as it were, always at the station. But it is not exactly clear what the system would cost.

Where money is available, great things can be done, whether now or in 2030. In San Francisco they have the Bay Area Rapid Transit (BART). Here space-age technology has produced a town transport system in which the stations resemble airport lounges and the trains are as comfortable and well-appointed as the most luxurious highspeed trains. All of them are driven by remote control by two computers. Tickets are issued, inspected and collected by machine. There are, indeed, attendants on each train but they are only there to drive the train should a computer break down or encounter a situation with which it has not been taught to cope. At busy times, the trains go faster. The system has cut commuting time in half.

An extension of this idea is the smaller rapid transit bus, capable of operating on demand, like a taxi, or in accordance with a timed schedule during peak travelling times. These electrically-powered vehicles are designed to function, like the BART, without drivers. They would carry a maximum of twenty passengers and, to avoid congestion, would operate on elevated monorails or guideways running alongside existing roads or railway lines. Computer-controlled circulation would ensure that collisions could not occur. At the busiest times of the day, the buses would call at specific stations to deposit and collect passengers. At other times, they could be summoned by pushing a button at any of the stations.

Is there no way to escape the cost of these elaborate systems of high-technology urban transport? Perhaps there is. Ordinary buses only use one quarter of the energy used by a car to carry one person, trams use half the amount and trains two-thirds. There are places – Holland is one – where the buses run punctually. If we could provide a bus service that was adequate, frequent and punctual, more motorists – whether today or fifty years hence – would be happy to leave their cars at home.

Long distance

Among the many bizarre schemes proposed to improve long-distance transport is one to provide an underground system that would enable a passenger to travel round the globe in a few hours, or across the North American continent in thirty minutes. Travelling in streamlined underground carriages, passengers would be carried through airless tunnels at over 22,500 kilometres an hour, propelled by electro-magnetic waves. Powerful magnetic fields with opposite polarities would cause the carriage to 'float' in the centre of the tube. Travelling in what is virtually a vacuum would minimize friction. The tunnels, besides carrying trains, would also be available for power and communications cables and slow-moving goods traffic. Short lengths by 2030? Perhaps. Round the world? I doubt it.

A slightly less ambitious scheme is based on the system that was once operated to carry cash and bills around large department stores by air pressure or vacuum. Cylindrical trains fitted tightly into three-metre diameter tubes would be propelled between cities by jet pumps that would both extract air from the tubes and then re-inject it at a higher pressure to blow the trains along like peas in a pea-shooter. One estimate puts the construction cost of a system like this as no more than that of a conventional motorway, which would carry the same volume of traffic. Again, the supporters claim that there would be no pollution and, once the system was built, no call for taking over large areas of agricultural land.

Train designers are already working on a new breed of monorail train that they hope will meet the needs of fast inter-city travel of the future. The existing system of wheels running on rails presents two major problems at speeds in excess of 400 kilometres an hour. Firstly there is insufficient friction between the wheels and the rails for propulsion and guidance and secondly, passengers travelling at this rate would be violently shaken. Since the proposed new trains will not have to be in contact with rails or anything else while they are in motion, and will hover by means of magnetic levitation, there will not be any vibration. The German firm, Krauss Maffei, has developed an experimental model that hovers above an elevated track in which is set an iron rail. Under the rail is an electromagnet, which, when a current is passed through it, is attracted to the rail. This magnetic force raises the train above the track and frees it from frictional contact. The train is propelled by a linear induction motor. This motor consists of two parts, the 'stator' and the 'rotor', which are able to move past each other with a small gap of air between them. The motor works through an alternating current passed through magnetic coils in one of the elements, setting up a magnetic field moving in a straight line. The moving field cuts the other element and this causes linear motion. To drive the train the magnetic field is induced in the track, giving a constant, smooth, noiseless thrust. The train can be stopped simply by reversing the magnetic field.

There are problems, however. For example, the gap between the magnet and the rail is only one or two centimetres. At speeds of over 300 kilometres an hour, the track would require constant maintenance to keep it smooth. Super-conducting magnets that could sustain gaps ten times as large and operate by repulsion rather than attraction would be mounted on the bottom of the train, and would cause eddy currents in the guide rails which, in turn, would set up a magnetic field with the same polarity as the magnets in the train. Since similar poles repel, the train would be levitated. These magnets would have to be supercooled in liquid helium so that their resistance to electric currents would virtually disappear.

Leaving a station, the train would be driven on rubber wheels by a conventional motor until it reached the 'take-off' speed of seventy kilometres per hour, at which point the magnetic levitation would be switched on and the train would leap forward down the track, silent except for the sound of the air rushing past its streamlined form.

These developments still have a very long way to go before we shall see railway companies seriously undertaking their complex operation. I very much doubt if there will be any major improvement on the wheel as a means of getting about on land by the year 2030. And while the wheel holds its place, so does the ship. With improvements in detail, the ship of 2030 will be a straight development of the ship of 1980. Aeroplanes may have taken away its passengers but it remains, as it has always been, the prime carrier of freight. One innovation we may see in 2030 as the price of fuel inexorably rises is the return of sail. Sail would not be the prime mover but it could become a valuable fuel-saving development. Similarly, in the development of air transport, we may see renewed interest in airships for the conveyance of freight. Since, following the disasters of the 1930s, the construction of airships ceased, major advances have occurred in plastic sheeting to cover the hull, in structional members to form the skeleton and in the lightness and performance of engines. A prophecy of there being airships in the sky in 2030 is a less hazardous bet than many more fashionable forecasts.

The Burning Issue

Power crazy

Stop anyone in the street today and ask them, 'What is energy?' and the chances are that most of them will reply 'oil', 'coal', or 'nuclear power'. Their conception of 'energy' will be directly correlated with its industrial and commercial applications. 'Energy' in their minds and in the minds of virtually everyone in the industrialized world has become synonymous with 'power'; domestic power, industrial power and, since the 1973 Middle East War, political power. Very few people asked this question would enter into a deep existential or metaphysical discussion on the meaning of 'energy'. Only a minority would try to explain that 'energy' is the mysterious, abstract ignition system of life, an infinite universal essence that causes the tides to rise and fall, the winds to blow and sends manic hares darting wildly across the downs in March. 'Energy' has moved from the poet William Blake's rosy state of 'eternal delight' and changed instead into an issue of anxiety and discord.

Einstein told the world that energy equalled mass multiplied by the velocity of light squared ($E=mc^2$). Politicians, the captains of industry and the media tell us energy is the life-blood of our industrial society, by which they mean the standard of living that we privileged members of the so-called developed nations have grown to expect. Remove energy and our lives would grind to a slow, inexorable halt. Prolonged power failures and descriptions of motorists in California brandishing firearms in petrol queues have helped to whip up public concern about energy consumption and resources, to the point where energy is now big news. Melancholy prognostications appear in the papers. We are told that the steady increase in the amount of energy used was level-pegging with world population growth from the time our prehistoric forefathers first set fire to a pile of twigs until 1900. During the twentieth century, however, the demand for energy has accelerated to such an extent that the amount consumed is doubling and then redoubling. The indications are that at the present rate of growth, if people go on doing in new circumstances what they have been doing in old ones, we shall be using twice as much energy in 2030 as we were using in 1970, giving a doubling period of only sixty years.

Thus the gloomiest forecasters have predicted that our existing resources of fossil fuels, coal, oil and natural gas will be exhausted by 2100. Of these the most emotive issue is oil. (Note: Although this is nonsense, I shall follow the popular idea that it raises. To fix an exhaustion date has little meaning because as fuel becomes more scarce and its price rises, more explorations are made, residues left behind in Texas oil wells become worth recovering, lower grades previously spurned are accepted. Oil wells do not run dry like petrol tanks, nor do coal mines become empty like coal scuttles. The process is a gradual one. Products become dear and scarce and people, who have survived deprivations before, use something else – or do without.)

The prophets of doom tell us that the remaining oil will have been used up in less than ten doubling periods, or under 100 years. Others claim with equal pessimism that the oil cartel of the OPEC countries will either severely restrict the flow of oil in the future or price it so high that the economic decline of the industrial nations will have set in long before the end of the twenty-first century.

One way and another, the writing is on the wall for those who want to believe the worst. For those disciples of the prophets of doom, the solution to the problem is perfectly clear. Other sources of energy must be found if our civilization is to survive; and with the emphasis on 'survive', nuclear energy is immediately rejected as being environmentally, socially and politically destructive. The idealized sources of energy that will power us through the twenty-first century into the bright new future must be non-polluting, harmless and infinite. In addition, to be effective in our salvation, they must be developed within the next fifty years.

Out in the noonday sun

With the possible exception of nuclear energy, and the heat of the earth's core, all the energy available to us originates ultimately from the sun. Our coal reserves were once the lush vegetation growing in the carboniferous forests of 300 million years ago. These forests owed their growth to photosynthesis for which the sun's rays provided the energy. The organic life that was transformed into pockets of oil and natural gas was similarly brought into being through solar energy.

Apart from bringing light, heat, rain and wind to the earth's surface, the sun is the giant 'power station' of our universe, generating far more power than we could ever need. In its constant process of converting, by nuclear fusion, hydrogen to helium, the sun changes vast quantities of matter into even greater amounts of energy. If it were possible for the earth to harness all this energy, every human being would individually have access to over 150,000 times the total power currently produced in western Europe.

Only a tiny fraction of the sun's full power reaches the earth. Of this, about 30% is reflected back into space; 47% is absorbed by the atmosphere, the lithosphere (land) and the hydrosphere (ocean); 23% is used in the hydrological cycle which brings rainfall and winds; and a minute proportion is used for photosynthesis to make plants grow.

In spite of the little amount we receive, the solar energy that does reach us is still impressive, by terrestial standards. Rays from the noonday sun falling from a cloudless tropical sky on to an area of eight square kilometres transmit as much energy as is generated by the whole of the CEGB supply system in Great Britain. On a larger scale, one day of sunlight falling on to the surface of Lake Erie in North America, is roughly equal to the entire energy needs of the USA for a year. With statistics like this to back them up, the supporters of solar energy may be forgiven for their optimism.

'Twinkle, twinkle little star,
How I wonder what you are!'

wrote Jane Taylor, an English matron contemporary with Jane Austen. A modern schoolboy would answer that a star was a quasar, a pulsar, a white-dwarf, a supernova, an asteroid or a meteor. Not so in the future though, for if scientists get their way and funds are forthcoming the answer to the question could also be that one star at least was an orbiting solar power station.

Given that much of the sun's power that reaches the outer atmosphere never filters through to us, it can be concluded that a solar-energy collector orbiting outside the earth's atmosphere would be far more efficient than collectors on its surface. In addition, a satellite placed in a suitable orbit would be in a position to receive sunlight nearly twenty-four hours a day. Maximum efficiency would be achieved by orbiting two of these satellites in space, 36,000 kilometres above the earth, so that one would be collecting a 'full load', while the other was on the dark side of the planet.

Each satellite would be fitted with large 'wings' covered with panels of solar-energy-receiving cells that would convert the sun's radiation into electricity, in the same way that existing photovoltaic cells did on Skylab. The electricity would then be fed into a control station where it would be converted into a microwave beam directed towards a receiving station on earth.

The receiving station would be constructed from antennae covering an area of 15·5 square kilometres. This would be large enough to collect all the power needed to satisfy the energy demand of a city the size of New York.

The supporters of these orbiting power stations point out that they would produce no pollution other than a small amount of heat; they would not rely on the earth's diminished resources of fossil fuels; they would be economical and safe; and they would encourage research into new forms of energy production. Their critics, on the other hand, are less optimistic. They argue that the construction of just one satellite would require the full commitment of NASA's space shuttle system to carry all the pieces of equipment into orbit. Numerous other flights would then be needed to assemble the power station and ferry the crews who would be needed to work them to and fro every few weeks. With our existing rocket equipment this might require one flight every hour for a century. These costs alone would be enormous, not to mention those of constructing the receiving station on earth. Contrary to predictions, the microwave beam would be a hazard to any living creature on which it fell. Apart from flocks of birds that might inadvertently fly through it, a deflection of just one degree could cast the beam on to the earth's surface more than 200 kilometres from its target.

When the orbiting power station was first proposed it was hoped that one might be in operation by the end of this century. Since the USA, the only western nation with the scale of technology required for such a mammoth undertaking, is struggling with an economic recession, it seems unlikely that significant advances will be made to make the orbiting power station a viable source of energy in the near future, if at all. An easier solution must lie nearer home.

The chief problem with harnessing the sunshine that falls on earth is that it is very spread out. This means that any system of collecting it would have to be spread over a very wide area as well. One scheme aims to use fields of silicon solar cells, which are cheaper to manufacture than those used in the current space programme. These silicon cells would produce direct electric current, which would need to be converted into alternating current before it could be fed into the existing electricity supply system. The area needed by these fields of cells is very large. One estimate claims that by the year 2040, the projected demands for electricity throughout the world would require fields of silicon cells collecting sunlight over an area larger than the whole of Yugoslavia.

The possibility of using a solar reflector has also been suggested. This is one of the oldest uses of solar power actively employed by man. Archimedes is claimed to have used giant mirrors made from hundreds of polished shields to set fire to the Roman fleet at Syracuse in 212 B.C. More recently, French scientists have constructed a solar furnace in the Pyrennees capable of generating a temperature of over 4000°C, which is hot enough to melt diamonds. Sixty-three giant mirrors with a total area of 2050 square metres of glass were spread over a hillside, reflecting the sun's rays on to a furnace 30 centimetres in diameter. As the sun moved across the sky, the mirrors turned with it to maintain a constant reflection on to the furnace.

If such a system were to be used to provide power to generate electricity on an industrial scale, the reflector system would need to be over 1250 times as large as the existing one in France and even then it would only generate power during daylight hours and on fine days.

This introduces one of the additional problems of solar power, the need to be able to store it for those periods when the sun does not shine. Existing batteries and accumulators are obviously inadequate. Major improvements will have to be made before it is possible to store enough electricity to power one domestic house, let alone an entire city.

Perhaps the greatest setback to the use of solar energy for the generation of electricity on a large scale is that the sun shines longest and most brightly on those parts of the world which are least developed and often least inhabited. In parts of the Sahara there are fewer than 100 hours of cloud the whole year. Yet there is nobody there to use all the sunshine and so far those people with the finance and the knowledge have hesitated to invest in an expensive scheme so far from home. Perhaps by 2030 they may have decided to do so: at least, there is plenty of room in the desert!

For those of us who live more than 45° north or south of the equator, solar power is of limited use. It can be employed to warm the bath-water and as a supplement to existing heating systems when the weather is fine. Unfortunately in Great Britain, on average, the sun only shines a little more than three and a half hours a day for half the year.

A touch of wind

If we could harness all the power of the winds blowing around the world we would have two and a half times as much energy as we require to produce all the power we needed. Estimates for the USA suggest that half the national electricity requirement could be generated by wind-power. Wind, like sunshine, is free, it is powerful, and it is plentiful. The trouble is that it is unreliable and altogether too widely dispersed.

Windmills have been used for 2000 years to move water and, more recently, to grind corn. Even before that wind had been used to move ships. Towards the end of the Second World War, Danish windmills were producing 3 million kilowatt hours of electricity a year, at a time when the country had very low reserves of oil and coal. In unindustrialized parts of the world, small windmills are used to drive the electric generators used to operate water pumps and, nowadays, to operate radio sets. Today, the possibility of using the wind as a major source of energy is being re-investigated.

One American plan aims at using the more powerful winds that blow over the sea to drive thousands of floating windmills moored off the east coast of North America. These windmills would be used to produce electricity by converting sea-water into hydrogen and oxygen. The hydrogen would be pumped ashore to fuel gas turbines which in turn would produce electricity. The main difficulty inherent in this scheme is the high cost of the windmills, at 50 billion dollars each.

Similar proposals have been put forward for land-based windmills which would stand 300 metres high and be driven by rotor blades 60 metres long. Past experience with giant windmills has not been very promising. The largest wind-powered generator ever built was constructed on a site in the State of Vermont. This colossal structure stood 30 metres high and was fitted with a pair of 8-ton, 53-metre diameter propellers. It generated enough electricity to light the nearby town for three and a half years until one of the propellers fell off during a storm.

To litter the landscape with big modern efficient windmills would not enhance the scenery, nor have windmills proved to be economical or reliable sources of energy.

Perhaps the best way to use wind power is to build sailing ships. Plans were already being drawn up in 1973 to build a 17,000-ton, 130-metre, square-rigged cargo ship in Germany. Unlike the old tea clippers, this modern *Cutty Sark* would be manned by a small crew whose primary function would be to operate the computer that controlled the sails. Instead of men having to scramble up the rattlings and inch their way gingerly along the yards to reef the sails, one man would control the whole intricate operation from the bridge. The designers estimated that the ship could reach a cruising speed of twenty knots, which would make her comparable with conventionally powered craft of the same size. She would be fitted with an auxiliary engine for manoeuvring in port or when becalmed.

The Canute complex

King Canute was probably the first recorded person to draw attention to the power of the tide and the waves, when he sat on his throne on the beach, with chilly water swirling round his ankles. Many tides have risen and fallen since Canute dried his feet and returned to his castle, but the practical application of the power of the tides has eluded all but a handful of men.

Tidal power has been used to drive water-wheels in a few suitable localities but, for the most part, the only energy derived from the action of the tide has been the use of combustible flotsom and jetsam washed up at high-water and gathered for firewood. Admittedly, the power of the world's tides is small compared with other possible sources of energy, such as nuclear power. But, even so, if it were possible to make use of all the tidal movements around the globe, there would be enough power generated to lift the entire human race to a height of three miles once every hour.

Unfortunately, tidal power raises problems similar to those presented by solar power. There is plenty of it, it will be available for as long as the earth and the moon survive, it is non-polluting and it is free. But it is troublesome to harness. Those who propose its widespread use look to existing tidal-power stations as examples of what future developments might look like.

The largest of these existing installations is situated across the estuary of the River Rance on the Atlantic coast of France. Here the tides rise and fall nearly 8·5 metres, and during the periods of spring tides the rate of flow through the turbines into the 22-kilometre deep storage basin reaches 18,000 cubic metres per second. When the tide turns, the ebb flows back through the turbines into the sea. At those times when the rate of flow is low the turbines act as pumps to raise the level of water in the basin. This pumping process increases the total electricity generated by half as much again for the expenditure of 10% of the total power output of the plant. As modern power stations go, the maximum output from Rance is small. Furthermore, due to the nature of the tides, it is only operational for a quarter of the time of a conventional power station operating twenty-four hours a day. This means that a large capital investment is needed to produce a relatively small return in terms of electrical output.

As, however, we come to accept that energy is no longer a cheap commodity, these disadvantages may be outweighed by the advantage of a free source of 'fuel'. Plans for tidal power stations in the Severn estuary and the Bay of Fundy, on the eastern seaboard of the USA, have been under debate for over fifty years. Reports from the USSR show that large tidal-power stations are planned for the Arctic coast, where tides rise and fall as much as 14 metres. The largest of these will generate an estimated 20 million kilowatts, more than thirty-six times the output of the Rance plant.

Whereas at least some progress has been achieved towards the possible use of tidal power, no practical system has yet been evolved to convert the motion of the waves into electricity.

Most of the projected schemes aim to use the up and down motion of the waves to pump sea-water through a turbine, thereby generating electricity. A simple proposal visualizes a 60-metre length of plastic pipe fitted with a check valve through which the water would be forced. Another more ambitious scheme involves a system of floating tanks anchored along a 1400-kilometre length of coast, pumping sea-water ashore to generate electricity. This, it is claimed, would be capable of generating half the total electricity requirement of the British Isles. However, the engineering and maintenance problems of such a scheme are formidable. The installation would present a danger to shipping, and the line of tanks would not improve the landscape.

One of the more plausible schemes aims to use the to and from motion of the waves to drive a series of vanes that would rock from side to side, generating electricity by their continual movement. The electricity would be used to convert sea-water to hydrogen and oxygen by electrolysis, as in the windmill scheme referred to before. This would overcome the need for energy storage in batteries since the gas could be conveniently stored in the unit until it was needed. The unit would be capable of moving from place to place. A group of 'rockers' would set out to sea under their own power. Once they were far enough from the shore, they would be turned abreast of the waves, and allowed to drift slowly back towards the coast with the vanes rocking to and fro, storing hydrogen as they went. When they reached the shore, the hydrogen would be drawn off from the tanks, enough being left to power the units out to sea again.

This scheme has received some support from the British government. So far, however, no practical outcome has been achieved.

Deep heat

The heat present deep down in the earth manifested in volcanic eruptions, hot-water springs and geysers has also been attracting attention from people looking for alternative energy supplies. British miners working in deep pits in Lancashire have to cope with high temperatures; the earth's temperature rises 1 °C for every 36 metres of descent.

At certain points, the earth's crust is thin enough to allow the molten magma to break through towards the surface, raising the temperature of the surrounding rocks and forming so-called 'hot-spots' beneath the surface. These 'hot-spots', or thermal reservoirs, are generally found in areas that lie along the edges of continental shelves and are frequently marked by the presence of volcanoes or geysers.

The estimates of the local geothermal power available from the earth's crust, at a depth of no more than ten kilometres, are very large. In the course of one year, it is claimed that 612×10^9 million megawatts of power could be generated. Compared with the 0·34 million megawatts consumed by the whole of the USA in a year, ten years ago, the potential scale of geothermal power becomes apparent.

Assuming that most of this would be untappable due to economic restrictions, even the fraction available to us would be adequate to make a useful contribution to the energy demands of the future.

Geothermal power is already being produced in areas where hot springs or geysers are common. In Hungary and Iceland, natural hot water is used for heating buildings within the immediate vicinity of its source. Underground steam is used to generate electricity in a few areas where there is easy access to the steam outlet. The Larderello power station, south of Florence, in Italy, has been generating electricity from geothermal power since 1904. Steam at a temperature of 240°C is tapped from reservoirs about 300 metres below ground. It passes through conventional steam turbines in thirteen adjacent plants to generate electricity for the surrounding neighbourhood. In New

Zealand the geysers at Wairakei are used to generate 8% of the national electricity supply.

However, as natural steam only occasionally occurs in places suitable for large-scale exploitation, underground hot water reservoirs are more likely possibilities for wide-scale use. The major problems with using hot water from natural sources are, paradoxically, that it is hot and natural. Since much of the water available for use is heated to temperatures in excess of 300°C, it dissolves a great many minerals, forming a highly concentrated solution of diverse chemical compounds. There are some hot waters extracted in California with a salt content of 20% compared with the sea which has a concentration of only 3%. Water of this type is extremely corrosive and a serious potential pollutant if released into surface waters. As a result, it can only be safely used in non-corrosive, closed, generating systems.

In these, the chemical brine is brought to the surface through bore holes and passed through a heat-exchanger system, where it transfers most of its heat to clean water flowing next to it, but in isolation from it. The clean water is boiled by the heat from the subterranean water and the steam produced is used to drive a conventional steam turbine. The brine is then passed through an extractor process that removes chemicals suspended in it before it is pumped back into its underground reservoir to be recycled.

A similar application of a heat-exchange system would allow the utilization of waters heated to lower temperatures, which constitute most of the underground water available in the reservoirs to which we could gain access. The heat-exchanger would have to be filled with a fluid with a low boiling point, since the temperature of much of the underground water would not be hot enough to boil water. Isobutane is one suggested fluid. This compound would behave similarly to the water system but at a lower temperature. The isobutane would boil and the vapour would be used to drive a turbine. The fluid would then be condensed into its liquid state to begin the cycle again.

Of all the geothermal resources available, the most attractive are the great beds of rock heated to temperatures of 300°–600°C which lie within three and a half kilometres of the earth's surface. The plans to make use of this energy require a hole to be bored into the stratum of hot rock to crack the rock, either by stimulating a subterranean explosion or by pouring acid into the hole. Once the initial cracks have been formed, surface water would be pumped into the hole to spread the cracks further into the hot rocks. The more water poured into the hole, the more extensive the network of cracks would become. Additional bore holes would be sunk at the same time to extract the water heated by the rocks. This water would emerge from the ground as steam which would either be fed into a turbine directly or passed through a heat-exchanger. Because water expands when it is heated, the column of cold water passing down the first bore hole into the hot rock is heavier than the heated water coming back up to the surface. It follows that the cycle, once started, operates through this difference in density with no necessity for mechanical pumping. Calculations made of the energy that might be expected from just over four cubic kilometres of hot rock indicate that it might be equivalent to the amount produced from 300 million barrels of oil. It would appear, therefore, that geothermal power could become an important source of energy in the future.

The concerted drive to investigate and develop all the alternative energy sources from the sun, the earth, air and water has been motivated by the belief that our existing resources of fossil fuels are going to disappear overnight. The simple answer to this is, as I commented earlier, that they will not. We are told that oil is going to run out, and that there will never be another Saudi Arabia. However, we find that there is oil in Alaska. It has also been found that there is oil at the South Pole. Now we find that the newly developed oil field in Mexico may be equal to that in Saudi Arabia. An alternative major potential energy source are the huge deposits of shale oil and tar sands for which exploitation techniques have not so far been worked out. How can we speak of shortage? The shortages of oil in recent years have been brought about by the increase in oil prices, not by the oil running out.

Conditions in 2030 will depend more on politics than on the natural resources of the ground. Besides oil there is coal. In Britain alone there have recently been discovered huge new deposits of coal sufficient to supply our needs into the twenty-third century, and nobody knows what further stocks of natural gas are to be found. Half of the continental shelves have still to be surveyed. Last of all is nuclear energy, which can provide unlimited power for those bold enough to take it.

Oil in the sand

The two principal deposits of tar sand are an area about the size of Belgium situated in the north of the province of Alberta in western Canada, and a much smaller deposit in eastern Venezuela which covers an area about the same size as the Grand Duchy of Luxembourg. Altogether it has been estimated that the world's tar sand deposits contain roughly 0·51% of its initial fossil-energy content. The amount available in Canada alone is thought to be double the amount of oil in the Arab oilfields.

Unlike conventional oil, the oil in the tar sands is a thick, bituminous substance, which does not flow in its natural state and consequently has to be mined. This mining operation is the main stumbling-block delaying the expansion of the tar-sand development. In the first place, the Canadian tar sands are located in a sparsely populated, inhospitable area. The land is covered with a spongy peat which makes transportation difficult. It is subject to bitterly cold winters and is plagued with flies in the summer. In addition, most of the tar sands are covered by glacial deposits to a depth of several hundred metres. This has first to be removed and dumped before the actual mining of the tar sand can begin.

The oil-bearing material is scooped out of the ground by

using the techniques of open-cast mining. It is then transported to a separation plant to be processed. Since about two tons of tar sand have to be mined to extract one barrel of oil, a plant aiming to produce 100,000 barrels of oil a day has to mine 200,000 tons of material each day.

In extracting the bitumen from the sand, the raw material is first mixed with hot water and steam to raise its temperature to 80°C. Then the mixture is filtered and the bitumen is sent to be upgraded to become high-quality crude oil. The hot sand is dumped back in the mine workings and the water is recycled.

Exploiting these resources will require a major investment programme to develop the technology that will be needed to overcome the problems of extracting the sand. It will be necessary, too, to develop a method of extracting the deepest deposits which account for two-thirds of the total resource. Sooner or later, however, and certainly before 2030, oil from these deposits will make a major contribution to energy supplies.

The Moses touch – oil from rocks

Throughout the world, there are deposits of sedimentary rocks laid down during different geological periods, which have been found to contain oil in the form of a solid material called kerogen. When this material is heated to about 370°C, the kerogen decomposes into a number of hydrocarbons some of which form a liquid similar to crude oil. These oil-bearing rocks are collectively referred to as oil shales. Oil from shale has been used in various parts of the world, from Italy, where it was used for street lighting, to the USSR, where it is used as a fuel in power stations.

Like tar sands, shale oils have to be mined on a large scale in order to make their extraction economic. Since much of the oil shales lie buried deep below the surface and are frequently interspersed with layers of rock bearing no oil at all, the cost of mining the kerogen often represents 60%

of the total cost of shale oil production. Disposal of the rock after the kerogen has been extracted presents a further problem, since during the processing it expands, so that it won't all go back into the hole from which it was dug.

The present mining methods are both costly and unsightly. This being so, consideration has been given to the possibility of extracting the oil from the rocks underground. The essential feature of such a technique would be to create cavities in the rocks into which the released oil could flow before being pumped to the surface. Subterranean fires fed by oxygen from the surface would be used to heat the kerogen. The difficulty lies in creating the cavities in the rocks in the first place and the suggestion has been made that this should be done by controlled nuclear explosions. Whatever method is eventually decided upon, a great deal of careful planning and costing will be necessary before the vast investment involved can be justified.

Nuclear energy – the final solution?

Conventional power stations that burn fossil fuels use large amounts of matter to produce relatively modest quantities of energy. In the production of nuclear energy, on the other hand, no combustion (which is a chemical process) takes place and a relatively small mass is converted, by a physical process of transmutation, into a huge amount of energy.

The fuel used in this process is the element uranium 235 (U-235) which is a radioactive isotope of the heavy metal uranium. In their natural state, isotopes like U-235 undergo radioactive decay, by which they give off energy and some of the particles of which their very substance is made up until such time as they have turned into a different material and have achieved a state of stability. All radioactive materials are potential sources of energy, but the rate of decay is so slow in many of them that they are unsuitable for the purposes of power production. U-235, however, displays a unique natural property; its atoms will occasionally split into two virtually identical fragments, as well as loosing off two or three neutron particles, together with an appreciable amount of energy. The nuclear industry began when it was discovered that the U-235 nucleus could be split by a 'collision' with a neutron travelling at the right speed. The first controlled nuclear chain-reaction took place on 2 December 1942, when the neutrons released from one 'collision' split a number of other nuclei, each of which released still more neutrons. The process had become a self-sustaining chain reaction giving out a steady flow of heat.

Since atoms are held together by large forces of energy, the amount of energy released when they are split is correspondingly large. In the case of U-235, the energy generated from the fission (splitting) of 500 grams of material (about a pound) is roughly equal to the amount produced from burning 1250 tons of coal.

Nuclear fission takes place in the following manner. The radioactive uranium, in the form of pellets, is placed in metal tubes 3·5 metres long. These tubes are packed into a thick-walled container which constitutes the core of the reactor. In the core, the rods are immersed in a coolant, which can be water, gas or any other suitable liquid. This slows down the bombardment of neutrons and regulates the fission of the uranium nuclei. The coolant ensures that the reactor will not overheat and melt the metal casing of the core. It is also the means of conveying the heat from the core to the heat-transfer system and thus producing the steam to drive the turbines to produce the electricity which is the object of the whole exercise.

The process is operated by moving a collection of rods filled with neutron-absorbing elements up or down in the core to control the movement of the neutrons and hence the output of energy. By lowering these rods right into the core, the fuel rods are isolated from each other, thus preventing the reaction from taking place and shutting down the reactor.

The big question with nuclear energy is of course safety. Although it is impossible for a nuclear reactor to explode like an atomic bomb, there is always a risk that radioactive material may leak out if, for example, coolant should escape, or if the core becomes so hot that it cracks, and the rods that would shut it down cannot go in properly. In 1957 radiation did leak from the plant at Calder Hall in Britain and in 1979 there was a breakdown at the Three Mile Island reactor in America. No one was hurt at the time in either incident.

Besides the problem of operating a nuclear power station safely, there is the added responsibility of disposing of used fuel. After fission has occurred, two kinds of by-products are produced. Both are radioactive and must be stored safely until their radioactivity has ceased. However, the half-life (that is the length of time for half the radioactivity to have been lost) of strontium 90 is twenty-eight years, that of caesium 137 is thirty years, while the radioactivity residual of plutonium 239 is virtually everlasting with a half-life of 25,000 years.

This means that used fuel has to be taken from the reactor to the re-processing plant in conditions of maximum safety. It is usually transported in large, sealed, cooled containers and is stored similarly after being re-processed.

The disposal of radioactive wastes obviously needs careful consideration. Underground storage in disused mine-workings is unsatisfactory since it is hard to insure that the highly corrosive material would never escape from its container and contaminate underground water supplies. Burial at sea presents a similar risk. It has been suggested that radioactive waste should be loaded into rockets and fired into the sun. A further idea is to solidify the material into glass-like blocks. This has been found to possess considerable merit.

Major contributions of energy have for a generation now been safely and successfully produced almost world-wide from nuclear power stations using nuclear fission. As has been described, there are problems and risks in the process. Nuclear fusion, on the other hand, would be an almost ideal solution to the world's energy problems. This is the process by which the sun generates its tremendous energy. The process on earth would take its fuel from sea-water and would produce next to no pollution, compared with nuclear fission. Instead of splitting heavy atoms like uranium 235, nuclear fusion involves the fusing (joining) of light atoms. These would be isotopes of hydrogen called deuterium and tritium and the metal lithium. In a process of controlled fusion one deuterium atom would be fused with one tritium atom and this would produce a helium isotope, helium 4, one released neutron and about 17·6 million volts of electrical energy.

With so much deuterium available in nature, the potential energy supplies are unlimited. The deuterium present in 1·5 cubic kilometres of sea-water used in a fusion reactor would produce as much energy as all the crude oil available on earth. Furthermore, the only radioactive element, tritium, has a half-life of twelve years, presenting negligible hazards compared with the wastes produced by nuclear fission.

Unfortunately in order to generate energy on the same principle as the sun, it is necessary to create similar conditions: a temperature of 100 M°C and an incredibly high pressure. When these conditions are successfully achieved, electrons separate from the nuclei and the gas separates into a thin conducting path, called plasma. Since there are no existing materials capable of withstanding such high temperatures and pressures, processes have to be developed that emable these two important conditions to be achieved. At present two methods are under experiment, but as yet neither condition has been attained, in spite of efforts extending over a number of years.

One of these processes employs the principal of magnetic containment. Experimenters have put the deuterium and tritium inside a torus, a doughnut-shaped space. Heat is then applied to electric coils wound round the torus. This is designed to raise the temperature of the gas to the fusion temperature. A second coil wound helically round the torus serves the purpose of keeping the plasma firmly in the centre by exerting a magnetic force that holds it in place, for were it allowed to touch the sides it would immediately cool and return to its state as a stable gas. At the moment of fusion a blanket of liquid lithium would be used to absorb free electrons and to transfer the heat generated to a heat-exchanger, to produce steam to drive the turbines. But, as I said above, this is merely theory, so far it has not been possible to create the temperature and the pressure needed to get fusion to take place.

The use of lasers would remove the need to generate these excessively high temperatures, since the laser itself is a controllable source of heat. The deuterium and tritium, compressed into pellet form, would be 'fired at' by a high-power laser beam. The shock induced by the laser beam would compress the nuclei together in the pellet for long enough, and under adequate pressure, for the fusion to take place. High-powered lasers could produce velocities in the pellets in excess of 3 million kilometres per hour, which would create pressures and temperatures similar to those inside the stars where fusion does take place. The neutrons shot out from this fusion would be trapped in a lithium blanket surrounding the fusion chamber, and this lithium would pass into a heat-exchanger.

Everything is ready, the techniques have been worked out. But so far nuclear fusion has only been achieved – uncontrolled – in the hydrogen bomb. We deserve success, we need success, we work for success, but so far it has eluded us. The prize is great indeed. My prophecy is bold. Provided we bend our efforts to the task, power from nuclear fusion will be available in 2030 and all our energy worries will be over.

The daisy complex

One day in 1780, an Italian housewife, one Signora Galvani, was standing at her kitchen sink preparing frogs for her husband's lunch and listening to the tutorial he was giving to a group of his medical students in the next room. As the conversation veered towards a particularly interesting part of the human anatomy, the lady's attention wandered from her task and she accidentally dropped the knife she was using. It fell on to one of the frogs which was lying on a zinc plate. To Signora Galvani's horror the dead reptile gave a violent jerk. She screamed and her husband dashed into the kitchen to see what disaster had befallen his wife. As a man of science he was understandably intrigued by her account of what had happened and in true empirical fashion he repeated the 'experiment'. The frog twitched again. Flushed with the prospect of fame and thankful for his good fortune, Signor Galvani returned to his students and modestly informed them that he had discovered animal electricity, the primary source of life. Of course he was wrong. We now know that what he had in fact discovered was the principal of thigh power.

'Daisy, Daisy, give me your answer do,' implores the young lover offering his sweetheart a lifetime of cycling, one behind the other, on a tandem. Now, if Daisy is a forward-thinking, environmentally-conscious young lady, she will realize the advantages of such an apparently unprepossessing proposal. Fossil fuels and horse fodder may run out, stranding the owners of carriages, both horsed and horseless, with no hope of using their sophisticated transport, but she and her beau will be able to travel as they have always travelled, by using four strong thighs.

The human thigh is a very efficient machine. It produces in work about 28%–30% of the energy contained in the fuel required to operate it, which puts it about the same level of efficiency as the conventional power stations that generate about twice as much 'waste heat' as they do electricity. Given the amount of capital that has been, and will continue to be, invested in these power stations, it does not seem unreasonable to consider the possibility of utilizing a vastly less complicated and cheaper source of power, based on a function with which we all have an intimate, personal knowledge.

In addition, since we already have extensive experience of handling all sorts of polymers, whether it is the natural polymer of rubber, which we can make artificially from petroleum, or whether it is nylon or perspex, there would appear to be no insuperable difficulty in manufacturing artificial thighs. These could be constructed from cables of artificial muscle fibre made from hydrocarbons, using the chemical constituents of real muscle as their basis. Once assembled into the complete 'muscle', the cables would be perfused with sugar, their fuel. All that would then be required would be a simple ignition system which would make them jerk tight and then relax, just as the tiny electric current passing between the steel knife and the zinc plate made Signor Galvani's lunch jerk, 200 years ago. The electric current required would be minimal, a mere pacemaker, and the only operational attention needed would be to insure a continuous sugar supply. Otherwise, the great 'muscles' would continue to contract and relax day in day out, driving a generator or whatever other machine they were attached to. In view of the fact that our own thigh muscles last as long as we do, the artificial polymer cables would only require occasional replacement, perhaps twice a century. No doubt to the sceptics this scheme sounds far-fetched, but is it any more so than the ideas of creating the conditions found in the stars?

Immediately after the last war, when Europe was prostrate, my cousin, Geoffrey Pyke, wrote an article in *The Economist* which advocated the use of thigh power as a solution to one of the most pressing problems at that time, the breakdown of the European economy. He argued that the destruction of the European communications system had brought the economy to its knees and had stranded the workers needed to put it back on its feet in displaced-persons camps dotted all over the continent. Now since each of these displaced persons, with some few exceptions, had two thighs, and since there was also an abundance of sugar, which for some reason nobody had thought to bomb, Geoffrey suggested putting these two together to create thigh power. He proposed that cheap cyclomotives should be built, each capable of holding forty or fifty displaced persons, sitting in ranks eight abreast. Once comfortably installed, they would proceed to pedal themselves across Europe, revitalizing their tired legs with occasional spoonfuls of sugar. Not only would the cyclomotives have solved the communications problem, they would have vividly demonstrated a means of transport that would have been a great deal more efficient than the steam locomotives that were eventually used. As it was, the cyclomotives were never built. But Geoffrey's principle is basically sound. For those who look upon the pending 'energy crisis' as tantamount to the state in which Europe found itself in the aftermath of the Second World War, thigh power offers a practical solution to the need for a cheap, non-polluting source of energy. It uses an easily renewable fuel and the engineering requirements lie within the bounds of our existing technology. So for anyone looking for future investments, let them put their money into polymer research with a view to synthesizing an artificial thigh which, when fuelled with sugar and caused to twitch, will constitute one, at least, of the power sources of 2030.

Daily Bread

Food, glorious food

The demographic predictions that there will be four times as many people on earth in 2030 as there were in 1970 may be right or wrong. World population is undoubtedly increasing, although not at the rate which simplistic arithmeticians predicted. But because human numbers are increasing, it does not follow that people are running out of food. The *kind* of food particular communities may eat can change but, throughout history, food supply has always kept pace with population numbers. It could even be argued that the increase in world population in modern times has been *because* of the increased wealth and the increased food supplies accompanying this prosperity. This being said, it is also true, as it has been throughout history, that some people do better than others. Modern technology has not affected the biblical understanding that 'the poor you have always with you', nor will it do so in the next fifty years.

In considering figures for calorie and protein consumption, it is important to remember that people can live at a high or at a low level of nutrition. Although prosperity is desirable and people, if they could, would choose to see their children plump, tall and forward for their age, hardship – unless it crosses the frontier into famine – does not necessarily kill. Indeed, I referred earlier to the evidence that some degree of hardship when young may be associated with a lengthened life span.

In reflection on food and health over the next fifty years and remembering how hard it is to deny that poverty and hardship have existed throughout history, it is salutary to remember that the total tonnage of available food world wide is influenced as much by likes and dislikes as by science and technology. Up to the present, a fundamental truth about human eating has been that people like meat. When they are prosperous, they eat more; when they are poor, they eat less.

The prosperous Americans eat, partly as the food they eat straight off the land and partly as the food their animals eat before they eat them, about eight times as much as poor people in Asia. Arithmetic says that if Americans chose to change their diet and eat bread and beans, the food they did not eat would feed undernourished populations elsewhere four times over. It is easy to calculate how big a tonnage of cereals could be grown on land used to grow fodder for animals and how many people could be fed if these cereals were shipped across the sea to feed them. It is also true that the protein a man obtains by eating meat is only 5–10% of the protein he would have obtained if he had eaten what the animal from which the meat came had eaten. It is also true that by planting one hectare of land with cereals, five times more protein can be produced than if the land was used to rear beef, and that this quantity could be doubled if the hectare grew peas or beans. If leafy vegetables were grown, the amount of protein produced would be fifteen times greater still. Nor would any one deny Georg Borgstrom's conclusion that the world's human population currently consumes only one-third of the vegetable protein consumed by its herds of cattle.

On the face of it, if we all became vegetarians tomorrow, the world could grow a tonnage of food several times greater than it does. Furthermore, Great Britain, which imports about half of its food, could become self-sufficient. Since the British import nearly three-quarters of the concentrated animal feeds they now need, they would save a lot of money if they stopped feeding livestock and started feeding themselves on vegetable protein. This sort of thing is still proposed by theoretical altruists. But I cannot help thinking it's silly. There surely is some sense in the community spending a reasonable amount of money on the food it likes, rather than putting it all in the bank. But perhaps our successors in 2030 will strike a better balance.

People enjoy eating meat and they will continue to enjoy eating meat fifty years from now. Indeed I believe that we shall be eating slightly more meat in 2030 than we do at present, because we shall be slightly richer, I believe, than we are today. The consumption of meat, I repeat, is closely linked with the relative wealth of the community. At present the British are less wealthy than they were. Consequently they are able to afford to eat less meat than they did, and some of what they are eating they spin out as Chinese take-aways and Indian food in which a little meat is expanded into a meal by being intermingled with other things. But when people become more affluent, they will start to buy more meat. Then the sandwiches, the Cornish pasties and the shepherd's pie will give way to the Sunday joint. Nor are they depriving the world's hungry. These they could help better by financial aid through, say, the World Bank, than by a self-conscious change of diet.

Meat, of course, has its merits in the nutritional scene. It contains amino acids in different proportions to vegetables. The average member of the human race – if there is such a person – obtains one-third of his or her protein from animal sources and about 10% of his or her calories. We ought also to remember that much of the vegetable protein now eaten by domestic animals could not be eaten by people. One of the best arguments for keeping ruminants such as cattle and sheep is that they can feed on grass and the like, which can be grown in areas unsuitable for the cultivation of any other crops, and which would be difficult to harvest and impossible to digest if we tried to eat them direct. Livestock thus actually create digestible high-quality protein from vegetable matter that would otherwise be of very limited use.

The usefulness of sheep and goats, when they are kept in their place, is even more striking. In upland areas, and arid regions, these are the only 'foodstuffs' that can be produced in difficult, often hilly, conditions and on poor soils. Remove them, and the sparsely vegetated upland and desert areas where they feed would become completely unproductive.

There was a time when pigs and poultry were used largely as scavengers, capable of eating waste products and clearing the land of old roots and vegetable debris, and giving in return meat and eggs. Prosperity and a more numerous population diminished the usefulness of such scavenging and, in response to the balance in the disposal of the money which prosperity brought, led to the introduction of more efficient ways of producing the bacon and eggs on which people were eager to spend their money. In place of the barnyard hen scratching in the dust, often broody and off her lay, came 300-eggs-a-year birds needing to be fed on high protein food, not so very different from that consumed by the farmer himself, although he would not thank you for saying so. It is only through the development of highly efficient methods of intensive farming that the variety and quality of food in industrialized communities, whose numbers have been steadily expanding from Europe to Japan, Korea and Taiwan in Asia, to South America, has become available.

Although food is both the instrument of nutrition as well as an economic commodity, desired for the satisfaction it brings by those who choose to buy it, there are those who, fixing their eyes on areas of scarcity, have given thought to alternative commodities that might be fit to eat. One of these is the cultivation of algae and yeast.

Algae are small plant-like organisms that live near the surface of water. They use sunlight to keep them going and, while growing, produce fats, starches and proteins. They absorb more energy from a given amount of sunlight than higher plants do and grow correspondingly quicker, often with a doubling rate measured in hours rather than weeks. Many algae have a higher concentration of protein in their tissues than conventional plants.

In certain areas on the fringes of the Sahara, one type of algae called *Spirulina platensis* is a regular source of food for the local population. The algae grow into a thick green scum on the surface of standing water at certain times of the year. This is skimmed from the surface to be cooked and eaten by the local people. Experimental units have been constructed in which to grow algae as a commercial agricultural operation. The algae are grown on tanks where they are exposed to the sun and fed with nitrates and phosphates. The carbon dioxide they need is obtained from the atmosphere. At harvest time, the liquid in the tanks is pumped through filters and the protein-rich material thus recovered.

Other schemes have been developed to harvest yeast as a food crop. Although yeast, unlike algae which obtain their sustenance from the carbon dioxide gas in the air through the intermediary of sunlight, must be fed, one scheme makes use of yeasts which are capable of growing on mineral oil. A crude oil-water mixture is supplemented with nitrogen and mineral salts.

The technical skill and scientific understanding required to develop workable schemes for the production of algae, of leaf protein – another potential source of protein – of yeast, and of single-cell protein from bacteria grown on natural gas on a commercial scale, have been impressive. It is, however, difficult to foresee whether by 2030 any significant tonnage of such material will be produced. The most attractive use for these products is as feed concentrates for pigs, poultry and cattle raised under intensive conditions. The high technology required to produce edible material from crude oil implies that such material is likely to be more expensive than the protein naturally provided by food crops, whether arable or animal. And the problem of sufficiency or shortage, as we have seen, is one of economics rather than technology.

Where dried yeast and S.C.P. (single cell protein) can compete with present concentrates, such as fish meal, they will find a market. Such may well occur by 2030 or sooner in a country such as Nigeria possessing an oil industry and the growing wealth that oil brings. Notions of such esoteric commodities filling a mythical 'protein gap' seem to me to be improbable. Nor does it seem at all likely that people will be found to eat much dried yeast, algae, leaf protein or SCP themselves. Apart from some hesitation in satisfying themselves about their wholesomeness, people are too conservative.

People's reluctance to accept new or unfamiliar foods is born out by history. Botanists have classified about 350,000 species of plants altogether. Of these, only about 3000 have ever been used for food, while 90% of our food comes from only a few dozen plant species.

Only three major new foods have been introduced into

our diet during the last 400 years. One of these is the potato, which Sir John Hawkins brought back from America in 1565. If he had arrived with it today, it would probably have been condemned by the food and drugs inspectors on the grounds that it contains a toxic ingredient, solanine, which would kill us if the potato contained five times as much of it as it actually does. Luckily, although the introduction of potatoes met some opposition, our ancestors were less fearful than we have become and potatoes have, therefore, served as a wholesome food for industrial populations for the last 400 years.

Maize is the second food, also introduced from the New World. Although the British mainly eat it as cornflakes, it is widely grown in other parts of Europe and is a major crop in the United States.

The third and most recent arrival is the soya bean. For centuries a staple food in China, its usefulness was only recognized in the United States a generation ago. It is now of major importance there.

At present the soya bean will not grow in Britain because the climate is too cool for it. Nor will it grow in the tropics because it needs days and nights of varying duration whereas near the Equator the nights and days are of the same length. Soya beans form a valuable crop. They are hardy; give big yields, and are an excellent source of good-quality protein. Because they are legumes, they enrich with nitrogen the land on which they are grown. They are an admirable food both for people and for livestock. Despite early setbacks, some progress has been made in developing TVP (textured vegetable protein) to be a substitute for meat in the same way that margarine is a substitute for butter. By 2030 this goal may have been attained.

But the most important prognostication for 2030 is that by then plant geneticists will have developed a strain of soya capable of being cultivated in the cool climates of Great Britain, Scandinavia and northern Europe where the most efficient agriculture is to be found. This would be of great value in increasing food production in these areas. More important still, if it is achieved, would be the development of soya strains capable of being grown in the tropics. If in those areas where today cassava and other starchy roots are the staple crop soya could be grown, public health would, at a stroke, be immeasurably improved.

Down on the farm

The chances are that the shape of farms and the character of the countryside will not alter much in the next fifty years. There is today a general feeling that although increase in size may increase efficiency, on balance the advantages of increased efficiency (even if such increase is brought about) is counterbalanced by damage to the quality of life. The argument has been going on for some time. During the Second World War, it was debated whether it was better to build large air-raid shelters in London or to issue each family with a small shelter to put in the back garden. For most nights, large strong shelters are better, for nobody gets killed; whereas with small shelters, every night some families become casualties. But on the one night when a bomb falls down the ventilator shaft in the big shelter, everybody is killed. Similarly in agriculture on a large scale, if the managing director of the 100,000-hectare ranch makes an error or absconds with the funds, everybody suffers.

If a week is a long time in politics, fifty years is too long altogether. It is, therefore, fruitless to speculate whether land nationalization or extended farm co-operatives will be the general mode in 2030. The problem of how to reduce the amount of capital immobilized in expensive machines required for only three weeks a year has already puzzled agricultural economists for more than fifty years and may continue to do so for another, so long as everybody needs to use his own expensive machine during the same three weeks.

By 2030, the rising cost and impending shortage of nitrogenous fertilizer needed to obtain maximum yields of farm crops may have been solved in three ways. I have just referred to the important possibilities inherent in any development of a soya bean variety capable of being grown in Great Britain. Like other legumes, soya beans are capable of 'fixing' nitrogen from the air, that is extracting it and returning it to the soil. This they do in collaboration with particular kinds of micro-organisms which grow on their roots. Research has been in progress to discover alternative types of micro-organisms capable of fixing nitrogen, not on the roots of legumes, but on those of cereal crops. If this were achieved – and I, for one, believe that it will be by 2030 – any shortage of nitrogenous fertilizer would be on the way to being brought to an end.

A third way by which any shortage of fertilizer would be met and its price likewise brought down, would be the solution of that problem – and how important it could be – of how to bring about nuclear fusion. Once this is done, we should have all the energy we need to fix all the nitrogen we want from the ample supplies forming 80% of the atmosphere.

There is, of course, another source of fertilizer which is, for several good reasons, neglected by industrial communities, namely sewage.

In many parts of the world human excrement is still used as a primary fertilizer. Less than 100 years ago, the 'night soil' men in China earned their living by collecting the 'night slops' from their neighbourhood every morning to spread on the fields. It is not without good reason that travellers to the Orient follow the wise maxim, 'No lettuces east of the Bosphorus.' Spreading untreated human excrement spreads cholera and other diseases apart from being smelly. The procedure used in China today is little different from what it used to be. Except within major cities, no significant quantity of human excrement is wasted.

An alternative procedure to spreading sewage on the land is to allow it to flow into ponds to feed fish, as described in the chapter on leisure. This, however, is only applicable on a small scale.

Even though the increased amount of leisure which urban citizens, at least, will be given in 2030 may cause some of them to want not only to visit farms but to work on them, the current trend towards agricultural automation is likely to continue. Large areas in, for example, the Mid-west of the USA, may well be cultivated and harvested entirely by machines operated from a remote control centre in place of expensive human workers. Professor Thring has suggested that one man might control several tractors or combine harvesters from a centrally located console, sending radio signals to the machines and following their progress on a television screen. Once in position to begin work on a field, the machines would be instructed gradually to work their way around or across it following a set pattern on a 'programmed map'. The tractor or combine harvester would be able to judge its position in the field by radio beams coming from each corner and would use photoelectric 'eyes' to 'see' the line of its previous circuit and steer itself as it negotiated the following one.

Another technological idea is that fields should be weeded by ultrasonic waves passed through the soil before the crop is planted. This process would be more efficient than the use of chemical weed-killers and would be free from their ecological disadvantages.

There's fish in the sea, no doubt of it

At present nearly all the fish caught throughout the world come from the waters over the continental shelves. This represents less than 10% of the area of the earth's oceans. Fishing here, however, has become highly efficient; large factory ships, belonging to the major fishing countries of the West, the eastern bloc and Japan, are capable of trawling deeper than conventional fishing vessels. A single Romanian ship in the Pacific is reported to have caught as many tons of fish in one day as the entire New Zealand fishing fleet.

It seems likely that all of the most productive fishing grounds are already being heavily fished and that the reason why the other areas of the ocean have remained untouched is that there is virtually nothing to be caught in them.

Despite these factors, the present level of the fish harvest could probably be increased, although careful steps would have to be enforced to avoid the risk of overfishing. Even if it were increasing, however, the quantity of fish caught in the sea would not keep pace with the present rate of increase in human population. Fish make an agreeable food and form a useful protein supplement to otherwise impoverished diets. But even if things could be organized so as to double the overall catch of fish by the end of the century, they would only supply an estimated 3% of the food needs of the projected world population, which is slightly less than they provided fifteen years ago.

By no means all of the fish caught are used for human food. In fact, about half of the total world catch is turned into fishmeal to be fed to livestock. One reason why this is so is dietetic conservatism, to which I have already referred. People like cod, herring, haddock and the other species to which they are accustomed. When less familiar fish appear, the public reaction is to convert them to the more popular and dietetically familiar bacon and eggs. Strange, isn't it? I am sure that, with only minor adjustment, we shall be doing the same thing in 2030. Some of the technological adjustments have already taken place. With the invention of the fish finger and the reduction in the numbers of herring and cod, less acceptable species have been incorporated into frozen units marketed acceptably under a coating of batter. Possibly in the future more unfamiliar species of fish will thus be rendered acceptable for mass consumption.

Old MacDonald had an aqualung

Mariculture, controlled marine farming, could do for fishing what agriculture did for hunting. That is, besides taking the danger and unpredictability out of it, making it more efficient, more productive and more profitable. It is, of course, unreasonable to dream of submarine hectares of giant kelp growing at the rate of a metre a day, since no use for it has yet been found, or of flocks of cod swimming the ocean like sheep. Codfish do not behave like this. There is, to be sure, virtue in the idea of farms in warm coastal waters, where fish could be reared in conditions that would best suit their growth. Perhaps the most convenient marine creatures to be farmed, other than oysters and mussels, which have been cultivated artificially for a long time, are lobsters. It has now been demonstrated that lobsters can be grown four times faster in artificially created environments than in the wild. They should be reared in water kept at a constant 23°C, which is heavily oxygenated by pumping air into their tanks. The lobsters are also encouraged to eat more than they would eat in their natural state. Under these conditions they reach maturity in two years, which is six years earlier than they do when they are living on the sea bottom. From specimens of these fast-growing lobsters, a fast-breeding strain has been developed which mature after only eighteen months. Lobster farms need to be situated near a supply of warm water, which could be the outflow of a power station, as long as no pollution is present.

Mussels and oysters have long been popular choices for intensive marine farming. Oysters particularly have been cultivated in many parts of the world. In the future, they are likely to be grown in net bags suspended across the sea bottom. As they mature, they need to be thinned out and attached to new ropes by nylon mesh. By this means enormous yields can be obtained. For example, an average of 300,000 kilos of crop per hectare of sea is confidentally expected, compared with 50 kilos of oysters or mussels per hectare for conventional oyster beds.

Eat, drink and be merry

With respect to restaurants, I see three kinds emerging in the next fifty years. The first is elaborate. The deep freeze contains meat and poultry, fruit out of season, fish of all sorts. Advanced technology will have replaced the kitchen porters and commis-chefs, the clutter of assistants and collaborators to be found in the kitchens of the *haute cuisine*, running errands for the chef and putting up the price of the meals. The devices for heating and browning, blending and emulsifying will be better than they are now and the food will be just as good: a marriage of aesthetics and scientific control.

The second type of restaurant will combine improved comfort with increased efficiency and will be a development of our present 'fast foods'. The articles will be standardized, but excellent. The selection will cover the most popular of Anglo/American, Italian, and Indian/Chinese. Sanitary control will be impeccable and the food will always be – when it is supposed to be – hot.

The third restaurant will be cheap. There will be bean-dishes, peasant mixtures from Africa, Asia, South America and the poorer states of the USA. The poor everywhere spend little on their food because they have little to spend. Yet, sitting round the cooking pot, they can enjoy their meals. I foresee a far-sighted entrepreneur, or a practical altruist, collecting such simple menus, adjusting them to taste, balancing their nutritional content, preparing them under the properly regulated conditions of food technology, selling them very cheaply (bring your own bowl and spoon) and enjoying immense popularity.

How will meals have changed by 2030? My forecast is: remarkably little. Breakfast, perhaps, may change as a social function in the next fifty years. The so-called 'cooked' breakfast has shown some tendency today in 1980 to wither away. Bacon, eggs, sausage and tomato is near to being a holiday treat. The busy citizen, conscious of what he believes to be the laws of health, is more likely to have a quick drink of orange juice, a light helping of breakfast cereal, a slice of toast – and away. If progress is made, in some small defence against unemployment, towards shorter working hours, the number of those starting work at 10.30 a.m. will grow. What more likely than that some original thinker, a setter of new fashion and custom, will start the custom of inviting friends to 'breakfast at 8'. Such an idea has real merit. The meal, though remaining simple, gives scope for elaboration. The guests will be fresh and unfatigued. Best of all, the duration of the party is self-limiting. Shortly after 10, the host and hostess know that most of those present must leave for work. And for those who do not enjoy alcoholic drink – and there are more of them than is often supposed – the gruesome necessity to suffer headache, nausea and confusion as the price of politeness is avoided.

What changes in foods and drinks ought we to expect? Who can say? Why do we only drink tea and coffee and have only done that during the last 200 years or so? Surely the field is open for a new flavour-sensation, some yet uninvented Coca-Cola, sharpened by some mild stimulant. There is an ingredient in cabbage-water believed to counteract hangover. Could this be the basis of a breakfast beverage? Once we drank ale, then chocolate, next coffee, then tea (except for the Chinese who started 4000 years before) – could cabbage-water come last in line? Since it's purely a matter of taste, I could be wrong.

Taste in food is unpredictable; who really expects to see through the fog to 2030? Take cornflakes. Who would imagine a religious food, conceived in the 1880s by the Seventh Day Adventists – pure food for the pure of heart, eaten untouched by human hands, unspoiled by cooking, straight from the box – would catch on, displace porridge, and become part of the morning diet across half the world?

It is extraordinary that we have given up eating chops and steaks for breakfast, being products of the sheep and the ox, yet continue to eat the flesh of the pig. Surely success lies ahead for the skilful technologist who cuts beef or mutton into thin wafers, preserves them in vinegar or pickles them in salt. Was the reason chemical, religious or aesthetic that left cured pork, as bacon, the winner of the breakfast steaks?

Another change in the diet of 2030 will be the final victory of margarine over butter. It will, I believe, be less of a victory than an Act of Union, like that between England and Scotland. Margarine has one great advantage over butter, namely, that it spreads straight from the fridge.

Dairy farmers have for 100 years, since margarine was invented, been trying to resist it, insisting that butter is a natural food. Now everyone knows it is impossible to go into a forest and find a kilo of butter hanging on a tree. Butter is the product of the high technology of the dairy industry. It is made in factories in which delicately controlled butter-making machines continuously churn the butterfat. In the future, when they make butter, they will make it to spread straight out of the fridge. This they will do by preparing it with some of the properties of margarine. On the other hand, margarine, in all its various qualities, will be made to possess – as many grades of it already do – all the agreeable properties of butter.

POSTSCRIPT TO THE FUTURE

The news of Monsieur Mège-Mouries's invention of margarine could have been greeted as a radical breakthrough in food technology when it was first made public. After all, it affected one of the most important functions of our daily lives, eating. Nevertheless, it has taken margarine more than a century to become a serious challenge to butter and even now it has not achieved absolute supremacy.

It would be foolhardy not to acknowledge that many similar scientific and technological developments will have taken place by the time the writers of 2030 sit down to look forward another fifty years to speculate on life in 2080. By that time cars may be running on hydrogen extracted from the ocean. We may all be living in automatically controlled homes and carrying portable telephones with us wherever we go. The world's energy crisis may have become history, with the successful application of nuclear fusion. We may be entertained by three-dimensional television, and take two-month holidays. Our education may last until the age of twenty-five and we may all be living for 120 years. Other, yet unimagined, changes may lie beyond the horizon waiting to take us all by surprise. Whatever form these changes take there is no denying the fact that the face of life in the future will be significantly changed.

Yet speaking as one who has the advantage of being able to look back more than fifty years, I look forward to the future with confidence because, rising above all the possible changes on the face of life, I foresee the dominance of the important human faculty of choice in determining the quality of our lives.

Looking back fifty years, we can see that what I call the paraphernalia of life has greatly changed. In 1930 there were many people in the world who had never seen a motor car or listened to a radio. A television set was still a technological dream. The moon was untroubled by man and his missiles, its back a mystery unseen since time began. Nuclear reactions existed only in theoretical physics and to fly the Atlantic from one continent to another was an adventure experienced by few. However, if we were suddenly transported back in time fifty years, I doubt if we would experience any great problems in settling into the pattern of the everyday life of 1930.

Similarly, it seems to me that fifty years from now people will choose to maintain the principles and values that they have always believed in. Material goods in the future may change the pattern of life, but the spirit which guides it will not alter. Monsieur Mège-Mouries's own nation may be right in their maxim, 'plus ça change, plus c'est la même chose'.

ROMANCE

☐ THE PROUD SERVANT	Margaret Irwin	£1.25
☐ THE GAY GALLIARD	Margaret Irwin	£1.25
☐ THE STRANGER PRINCE	Margaret Irwin	£1.25
☐ SO WICKED MY DESIRE	Stephanie Blake	£1.50
☐ MYSTIC ROSE	Patricia Gallagher	£1.20
☐ CAPTIVE BRIDE	Johanna Lindsey	£1.00
☐ A PIRATE'S LOVE	Johanna Lindsey	£1.20
☐ GILLIANE	Roberta Gellis	£1.50
☐ GREEN GIRL	Sandra Heath	80p
☐ THE TALL ONE	Barbara Jefferis	£1.00
☐ ROYAL MISTRESS	Patricia Campbell Horton	£1.50

SCIENCE FICTION

☐ THE OTHER LOG OF PHILEAS FOGG	Philip José Farmer	80p
☐ GRIMM'S WORLD	Vernor Vinge	75p
☐ A TOUCH OF STRANGE	Theodore Sturgeon	85p
☐ THE SILENT INVADERS	Robert Silverberg	80p
☐ THE SEED OF EARTH	Robert Silverberg	80p
☐ CRITICAL THRESHOLD	Brian M. Stableford	75p
☐ THE FLORIANS	Brian M. Stableford	80p
☐ FURY	Henry Kuttner	80p
☐ MUTANT	Henry Kuttner	90p
☐ HEALER	F. Paul Wilson	80p
☐ JOURNEY	Marta Randall	£1.00
☐ THE GAMEPLAYERS OF ZAN	M. A. Foster	£1.25
☐ PROJECT BARRIER	Daniel F. Galouye	80p
☐ THE MIND THING	Fredric Brown	90p
☐ THE PROUD ENEMY	F. M. Busby	90p

HORROR/OCCULT

☐ TRANCE	Joy Fielding	90p
☐ DEVIL'S COACH-HORSE	Richard Lewis	85p
☐ ISOBEL	Jane Parkhurst	£1.00
☐ RETURN OF THE HOWLING	Gary Brandner	85p
☐ RETURN OF THE LIVING DEAD	John Russo	80p
☐ DYING LIGHT	Evan Chandler	85p
☐ RATTLERS	Joseph L. Gilmore	85p
☐ LOCUSTS	Guy N. Smith	85p

FILM/TV TIE IN

☐ WUTHERING HEIGHTS	Emily Brontë	80p
☐ STAND ON IT	Stroker Ace	95p
☐ THE ROSE MEDALLION	James Grant	90p
☐ AMERICAN GIGOLO	Timothy Harris	90p

(H13B:10–12:79)